FIBER TO THE HOME

WILEY SURVIVAL GUIDES IN ENGINEERING AND SCIENCE

Emmanuel Desurvire, Editor

FIBER TO THE HOME

The New Empowerment

Paul E. Green, Jr.

WILEY-INTERSCIENCE

A John Wiley & Sons, Inc., Publication

Published by John Wiley & Sons, Inc., Hoboken, New Jersey
Published simultaneously in Canada

For general information on our other products and services or for technical support, please contact our
Customer Care Department within the United States at (800) 762-2974, outside the United States at
(317) 572-3993 or fax (317) 572-4002.

Wiley also publishes its books in a variety of electronic formats. Some content that appears in print
may not be available in electronic formats. For more information about Wiley products, visit our web site
at www.wiley.com.

Library of Congress Cataloging-in-Publication Data:

Green, Paul Eliot, 1924–
    Fiber to the home : the new empowerment / by Paul E. Green, Jr.
        p. cm.
    "A Wiley-Interscience publication."
    Includes bibliographical references and index.
    ISBN-13: 978-0-471-74247-0
    ISBN-10: 0-471-74247-3
        1. Optical fiber subscriber loops. I. Title

    TK5103.592.O68G74 2006
    004.6'4–dc22   2005048607

Printed in the United States of America

10   9   8   7   6   5   4   3   2   1

Contents

CHAPTER 2
Architectures and Standards, 27

CHAPTER 3
Base Technologies, 69

Foreword

The promise of fiber communications has not been realized either in the long-haul backbone network or in the last mile access network. Considerable discussion has taken place regarding the failure of the former, the most compelling explanation being the bursting of the dot.com bubble which drastically reduced the growth of traffic in the backbone and produced a devastating effect on the telecomm industry. Capital for broadband in the backbone dried up rapidly at the turn of this century.

More complex reasons exist that fiber to the home (FTTH) has not taken off in the past. In this book, Paul Green examines this issue and provides an in depth treatment of the drivers that are now emerging which will likely spur dramatic increases in FTTH deployment (already the early signs are there) as a strong alternative to the existing broadband access technologies of copper (DSL) and coax (cable modems). He makes the effective, and common, argument that today's move toward multimedia streams into the home (in addition to voice and data, the three together forming the holy grail of the "triple play") is driving the demand for broadband access to homes (as well as other end–user premises). Green minimizes the value of copper and coax as a sustainable solution while at the same time arguing that fiber is future proof, and hence the correct solution. The influence of the common carriers, especially in the United States, on the continued push for DSL and of the cable operators on the continued push for cable modems is an important part of the discussion in which Green engages us.

Paul Green takes us on a journey through all aspects of the FTTH landscape. He has crafted an exceptional book that explains why Fiber to the Home is finally coming to your neighbourhood. After arguing, most effectively, that the access network represents an enormous bandwidth gap between the backbone network and the end user computational platforms, he then takes us through the many layers of design and device issues related to fiber.

First we are exposed to the different architectural choices for passive optical networks (PONs) leading finally to a very nice summary table that compares the ATM-based and Ethernet-based PONs; he then offers his opinion that "...the most striking difference, in this modern world of IP packets, the Web and ubiquitous small, cheap laptops and desktops, is the complexity of the (ATM-based) APONs compared to the (Ethernet-based) EPONs... much is due to the tyranny of the 125-microsecond framing" that comes from the telco world. He further offers "It is this author's prediction that EPONs and their descendants are likely to become the norm everywhere" and he further slams the ATM-based technologies with "...the heritage of ATM cells and 125-microsecond framing are increasingly likely to be seen as an expensive luxury from the past, eventually achieving only a lingering archeological significance".

Next, we are presented with the base technologies underlying FTTH and Green quickly observes that "...the principle technology challenge of FTTH has been to cost-reduce the appropriate subset of historically available solutions, rather than to invent new ones". He lays down the physics underpinnings of the various elements that make up the FTTH system in a straightforward fashion that is eminently understandable. In so doing, he lifts the veil of mystery about how fiber optics works.

We then return to the more practical issues of deploying FTTH systems, after which I felt the urge to purchase a hardhat and begin installing fiber in my neighborhood. The state of current deployments worldwide is a most interesting chapter as well, where we are treated to a comparison of the many different drivers and levels of penetration across the world. Sadly, it is the case that the United States is sixteenth in terms of broadband penetration on a percentage basis, with many of the Asian countries way ahead of the United States.

In summary, fiber is finally making its move into the edge of the network. The technology is sound, the demand is here, and the cost structure is compelling. Paul Green has been working in fiber optics for decades and has been a champion of the tremendous bandwidth and low attenuation offered by something so thin and so small. His deep expertise in fiber optics specifically, and digital communications in general comes through in this excellent book. If you want to know all things about FTTH, you have picked the right book.

LEONARD KLEINROCK
Professor of Computer Science
UCLA

Los Angeles, California
August, 2005

Preface

Ever since the days of the caveman and his tom-toms, humans have used a bewildering variety of communication technologies: smoke signals, flags, copper wire pairs, coax, two-way or broadcast terrestrial and satellite radio, carrier-current over power lines, point–point microwave, free-space infrared, and, of course, optical fiber. Out of this grab bag of delivery mechanisms, two stand out as likely to dominate all others as the twenty-first century unfolds: radio (wireless) for ubiquitous coverage with limited capacity and fiber for almost unlimited capacity but highly constrained coverage.

This book is about the latter of these enduring options; actually about one rapidly evolving aspect of the latter, fiber to the home. When we speak of "fiber to the home" (FTTH), we mean to include also fiber to the user's premises (FTTP) and fiber to the business (FTTB)—any situation where the optical path in the fiber runs all the way to individual end-user premises. This might be a single residence, single multidwelling unit (MDU), or a single business establishment, and the fiber path might enter the building or terminate elsewhere on the premises. What is not included in FTTH is any situation where electronics are interposed along the path from central "head end" to user's premises. In other words, we deal here with OOO situations, not OEO—with a form of "all optical network" (O for optical, E for electronic).

Fiber is already well entrenched as the medium of choice for intercontinental and intercity communication—in other words the "backbone" or "interoffice" and "metro" segments. That battle for an optimum solution has been won. It is the promise and the methodology of extending fiber onward to the user's premises that is the current battle and the subject of this book. The increasingly compelling argument that existing and proposed digital subscriber line (DSL) and cable modems are merely temporary and insufficient fixes for the "broadband last mile" is based on their limited capacity, limited reliability, large lifetime costs, and their lack of future-proofness.

So, how badly do we need FTTH *now*? Arrayed on the side of "getting on with it" are the ever increasing user bandwidth requirements, competitive pressures on DSL from cable, missed opportunities in the computer industry that depend on high bit rates, falling technology costs, maturity of the architectures and standards, and issues of international industrial competitiveness in a world where an enduring broadband infrastructure is seen as an important national asset. Such positive forces are opposed by very strong pressures "not to get on with it": the immense cost of replacing copper, the need for uninterrupted premises equipment powering, conservative and short-term business strategies that force unnatural extensions of the service lifetime of the stranded investment in copper, an entrenched habit of dealing with restrictive legacy regulatory entanglements by means of endless lawsuits, and in some countries (notably the United States) an almost total absence of any effective government encouragement.

We shall discuss all these factors in this book. We first set the stage by reviewing the many needs of society in general and individual users in particular for delivered performance parameters beyond the capabilities of DSL, wireless, and cable. We then discuss the candidate architectures or "block diagrams" of FTTH systems and describe the existing standards. Next comes a discussion of the technologies involved (optical fibers, splitters, amplifiers, filters, and electrooptical terminal equipment), and then the methods of deploying them. Then we present a snapshot of the significant current actual deployment worldwide. We close by speculating about the future evolution of the matters discussed in the earlier chapters.

This is a rapidly evolving and multidisciplinary field, and one of suddenly growing importance, as the inadequacies of past solutions and competitive and regulatory pressures for new solutions are finally sinking in and action is being taken. It is a field that many engineers, enterpreneurs, technicians, and planners now need to know something about. In this spirit, and in the spirit of the whole Wiley Survival Guide Series, the intent here is to get the reader satisfactorily up to speed, all in one volume. To facilitate this process, each chapter concludes with a little Vocabulary Quiz, a list of the topics and terminology with which the person new to this field will want to feel comfortable. For easy cross-referencing, all such terms appear in Boldface in the text.

I would like to thank Sam Greenholtz, Mike Kunigonis, Mike Render, and Bob Whitman for educating me and for help in locating material for this volume.

PAUL E. GREEN, JR.

Mount Kisco, New York
September, 2005

CHAPTER 1

Evolution of the Broadband Last Mile

■ 1.1 INTRODUCTION

There are some technological plateaus that can be glimpsed from a great distance and forecast fairly accurately but that come so clearly into view as to gain immediacy and irresistibility only after a journey of many years. Such an end point is the one discussed in this book, the replacement of copper by optical fiber for fixed communication across the "last mile," the so-called access component of the telecommunications infrastructure. The completion of the evolving fiber infrastructure to include the access links has been forecast for years. However, until recently the various tentative engineering studies, field demonstrations, and other gestures toward an all-fiber future were succeeded by very little follow-through. Now **fiber to the home (FTTH)** is happening. And it is happening at many places around the globe.

A combination of circumstances has caused this change. The growing bandwidth needs of individual users on both sides of the economic "digital divide" have combined with the development needs of individual communities and indeed entire nations to constitute a considerable user pressure. Quantitatively, these pressures for more ease of use and more bandwidth to more users are rapidly exceeding the capability of even the most highly evolved copper-based or wireless solutions. This bandwidth bottleneck has been understood for years. What has perhaps been the most immediate new triggering event for today's long-awaited FTTH movement, at least in the United States, has been the threat to the continued survival of the established local access carriers posed by competition from cable technology and, to a lesser extent, from cellular radio.

These are some of the things that are *pushing* the movement toward FTTH, and we shall spend the rest of this chapter dealing with them. But there are also things

Fiber to the Home: The New Empowerment, by Paul E. Green, Jr.
Copyright © 2006 John Wiley & Sons, Inc.

that are *pulling* people toward such solutions, and these are the thrust of the other chapters in this book. Stable and exhaustively standardized architectures now exist, forming a set of blueprints from which the systems can be more easily and accurately implemented. The technology piece-parts are evolving steadily in cost-effectiveness. A clever set of innovations in fiber installation techniques has made what once was the most expensive part of the undertaking much more economical. And one sees from a worldwide examination of the progress of FTTH that governments at both state and national levels are finding ways to encourage all the players to move ahead with this long-postponed but inevitable development.

◼ 1.2 A FEW DEFINITIONS

Fiber's properties are so uniquely suited to this job that substitution of fiber for copper in the broadband last mile, or **access**, may very well be essentially permanent. The access is the portion between the central office or cable head end and the subscriber's premises and is the only part of the fixed telecommunications plant that has not already evolved completely to fiber. Fiber has already become the dominant medium in the **metro** (tens of kilometers) and the **interoffice (IOF)** or **long haul** (hundreds and thousands of kilometers) but not as yet in the access. In fact, the innovations of **dense wavelength division multiplexing (DWDM)** [Green, Ramaswami] have resulted in so much overcapacity in both the metro and long-haul arenas that the market for high-density WDM has more or less flattened out. But its growth will return, propelled in great measure by the last-mile technologies, new forms of media content, and the societal pressures sketched in this book.

So, fiber has become the preferred paving material for all the high-capacity telecomm highways that span continental and intercontinental distances interconnecting the powerful switching centers, telephone exchanges, and cable head ends with each other. But why exactly is it written in the stars that it will also be fiber that replaces twisted-pair copper and coaxial copper in the infamous last mile by which these centers communicate with individual user premises? Why would one need all that bandwidth for such a localized communication need? And why is not the access job to be done adequately by a mix of copper and *untethered* forms of communications, notably cellular wireless, whose progress has recently been so spectacular, while that of the fiber last mile has been so unspectacular?

In point of fact, there have always been things that copper and radio cannot do but that fiber can. They just have not been that important for the access environment until now. They have to do with bandwidth capacity certainly [e.g., the capacity to carry multiple **high-definition television (HDTV)** channels], but there is more to it than just bandwidth. There are matters of signal attenuation, first cost, lifetime serviceability costs, convenience to the user, and the business advantages of combining all the communication services into one bundle, one management process, and one monthly customer bill.

It has long been realized that copper simply cannot suffice to bridge the enormous and growing capacity gap between our desktops, laptops, or TVs and the equally enormous and growing capacity that this very medium, fiber, has brought to metro and interoffice facilities. This **last-mile bottleneck** is partly quantified in Figure 1.1. In the interoffice environment of both telcos and cable providers, raw bit rates of 2.5 to 10 Gb/s *per wavelength* have been standard for 5 years or more. At the user end, 2 to 8 Gb/s data transfer rates within the ubiquitous laptops and desktops are common and this number is growing. [While a full 32 Gb/s might flow between memory and central processing unit (CPU), at least 1 Gb/s can flow to and from peripherals.] Some predict that data transfers inside personal computers (PCs) and laptops will reach 100 Gb/s by 2010. For HDTV offerings to expand into hundreds of channels, as has happened with standard definition TV, will require several gigabits per second (assuming tens of megabits/second per channel).

So there is *already* a bandwidth bottleneck of at least 3 to 4 orders of magnitude between the communications supplier and his customers, in spite of aggressive use of copper-based last-mile technologies by the telcos and cable providers. Moreover, the high degree of capacity asymmetry dictated by both **DSL (digital subscriber line)** and cable technologies is counter to the trend toward more peer-symmetric traffic loads for some important new applications. It should also be emphasized that the bit rates available from fiber are almost distance independent over the usual span of up to 20 km access distance, whereas attenuation in copper poses many serious systems problems even at these distances, as we shall see.

One concludes that while DSL and cable modems are interesting and involve some cunning and interesting technology innovations, they hardly represent an

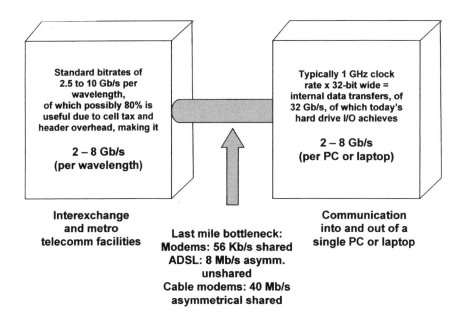

FIGURE 1.1 "Last-mile bottleneck" between a user's desktop or laptop computer and the metro and long-haul facilities.

adequate response to the user demand growth we shall discuss next, and even less support a game-changing revolution called for by the present situation. We shall quantify the inadequacies of cable and especially DSL in Section 1.12.

Figure 1.2 shows that to serve a significant fraction of customers in places such as North America requires a technology capable of distances of tens of kilometers. In some developing countries the distance requirements are even more severe. As we shall see later in this chapter, for today's interesting bit rates this is leading to some strange copper–fiber hybrids. The permanent solution is clearly to get rid of the copper.

Figure 1.3 shows the two most common schemes by which fiber is used to connect user premises (home, businesses, apartment buildings) to the central offices of the phone companies or the cable companies optically. For larger businesses, ring connection by fiber is also used, so that, in case a fiber is cut, **automatic protection switching (APS)** acts to send the traffic backwards around the ring to its intended destination.

In the **star** or **direct fiber** or **homerun** connection, there is one port at the central office for each subscriber, whereas with the **passive optical network (PON)** there is one for every N, N being typically somewhere between 8 and 64. These designations are often abbreviated **P2P** (point-to-point) and **P2MP** (point-to-multipoint), respectively. The PON architecture and set of signaling protocols originated in the cable industry [Ransom] where the one-to-many nature of the topology was almost exactly the copper equivalent of the fiber version shown in Figure 1.3(*b*). Because of its economies, attributable to the use of time slicing of traffic in both directions, the PON alternative for FTTH is being much more

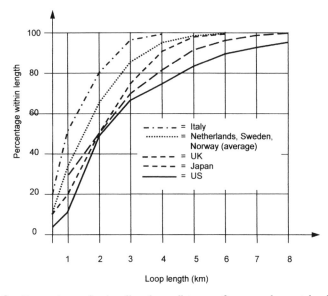

FIGURE 1.2 Comparison of subscriber loop distances for several countries [cumulative fraction of central office (CO)-to-subscriber distances less than a given number of kilometers] [Mickelson].

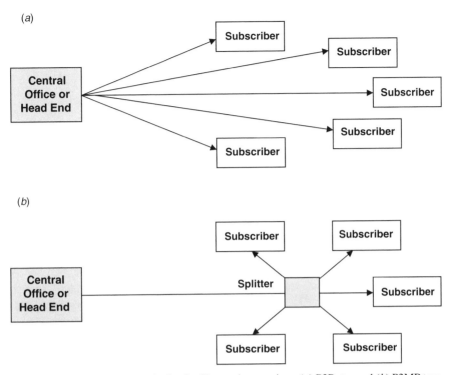

FIGURE 1.3 Two basic topologies for fiber to the premises. (*a*) P2P star and (*b*) P2MP tree, or passive optical network (PON).

widely installed today than is the star connection. Because of this dominance of the PON alternative, the treatments in this book will be spending most of the time with that configuration. Eventually, when per-user bandwidth needs grow to levels that cannot be supported by timesharing the fiber, the star connection will be mandated, but that is not expected for many years.

The optoelectronic element at the head end is universally referred to as the **optical line terminal (OLT)**, but until recently a variety of names was given to the corresponding element at the subscriber end. One will see the names **network interface device (NID), optical network terminal (ONT)**, and **optical network unit (ONU)** applied to what is essentially the same piece of equipment. Usage is tending toward the use of ONU as the preferred term. This is the term we shall use.

In this chapter we begin by examining in some detail the pressures that underlie the long-awaited fiber last mile. There are competitive pressures on telephone companies from cable companies. There are pressures from end users for support of more and better high-speed applications and for lower costs. And there are pressures from both the telecomm industry and the information industry for new kinds of revenue-producing services and products. After reviewing these matters, we then briefly discuss the many reasons that today's legacy broadband last-mile technologies—cable, DSL, cellular and satellite—are insufficient to meet these challenges. For completeness, we also summarize the position of the skeptics, who argue, for example, that, while fiber access is an elegant solution and fine for large businesses, most of the end-user world will not need it for a very long time.

■ 1.3 CABLE COMPETITION

The providers of cable television, often called **multiple service offerers (MSOs)**, have taken the lead in providing broadband services. (Since recent regulatory relief has provided the telcos too with the opportunity to provide multiple services, the term MSO to refer only to cable providers has become obsolete and will not be used here.) Provision of cable services over and above broadcast television has evolved to encompass also digital TV as well as bidirectional data and voice. Voice, or **POTS (plain old telephone service)**, has recently evolved into **voice-over Internet Protocol (VoIP)** digital bit streams. As we shall see later in this chapter, technology evolution has dealt the coax-based cable providers a better hand for supplying broadband than is available to phone companies who are tempted to base their solutions on the traditional copper twisted pair.

A measure of the U.S. phone companies' current level of distress can be seen in Figure 1.4, which shows the sharp downturn in the traditional voice-grade cash cow, POTS, due to competitive losses to cellular and to VoIP offered by cable companies and competitive local exchange carriers. Because the cable providers have a better technology base than do the telcos, are less encumbered by regulatory restrictions, and have traditionally displayed a more entrepreneurial mindset, their recent actions

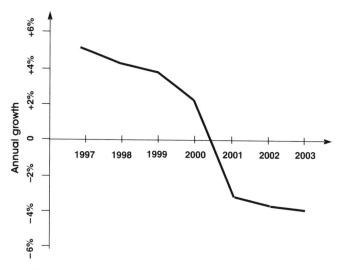

FIGURE 1.4 Collapse of the U.S. POTS business—years of growth followed by years of loss to cable and VoIP [Bernstein].

have had the effect, at least in the United States, of causing those telephone companies that depend on local access for much of their revenue to respond by looking for a technology that is at least as good as coax and hopefully better. They have not found it in cellular wireless because of coverage, capacity, and reliability limitations, but by their recent moves toward fiber to the curb, premises, and home, they have begun creating the inevitable bimodal future: wireless for reachability, fiber for capacity.

Some industry observers [e.g., Ferguson] have argued that the cable industry has become as noncompetitive and monopolistic as the telco industry, thus forming a duopoly, with "gentlemen's agreements" not only within but also between the two halves not to compete. In the telco half, there appear to have been written or unwritten agreements among the large **incumbent local exchange carriers (ILECs)** not to invade each other's territory. However, to the extent that telecomm in the United States has been a duopoly, it seems to be less true with every passing month. To this observer, it appears that not only has competition between the two monopolies intensified but so has competition within the telco half. The recent competitive race by successful ILECs to buy out not only cable franchises but especially some unsuccessful long-haul carriers is leading to a situation where traditional geographic turf boundaries between ILECs and between telcos and cable providers are being blurred or are disappearing altogether.

■ 1.4 TRIPLE PLAY

As a solution to these user pressures, the favored service offering for broadband today, whether DSL, cable, or FTTx, is a standard mix of traffic types referred to as **triple play**. The three traffic components are bidirectional voice bit streams, video (unidirectional and usually analog up to now), and bidirectional high-speed data. Today it may be difficult to imagine why this categorization does not cover every foreseeable contingency, but if new classes or subclasses of traffic evolve, the huge bandwidth and the protocol and format transparency of fiber will accommodate them gracefully. It seems likely that the first addition to the present mix will be to make the video component bidirectional.

The introduction of triple-play offerings by the U.S. cable companies has recently raised the competitive stakes to a level that begins to appear intolerable to the incumbent common carriers. As we shall discuss in Chapter 5, while triple play may be the standard emerging broadband offering in North America, in some other parts of the world, regulatory restrictions mean that not all three components can be carried by the same facilities.

■ 1.5 INTERNATIONAL COMPETITION

There is not only the issue of competition between industry sectors within one country but also that of the relative development of the infrastructures of various nations as they compete for industrial success in the rest of the twenty-first century. There is very little that broadband in general and FTTH in particular can do as yet to help the plight of developing countries at this early point in time, but developed countries present a different picture. As we shall see in Chapter 5, the nations of the Far East have far outpaced those of North America and Europe in their embrace of this new technology. Almost every government in Asia seems to see a very high capacity, future-proof telecommunication infrastructure as a necessary national resource for its industries and citizens and has proceeded accordingly. Japan, where distances are short enough to tempt the judgment that DSL copper is sufficient, nevertheless leads the world in per-capita availability of fiber to the premises. By contrast, the United States, even though its loop distances are longer, nevertheless has a tradition that the government must not play favorites with a particular technology or industry sector (with the exception of defense). Therefore, this opportunity for infrastructure upgrade has been dealt with by leaving it up to what has until recently been only a collection of small companies in smaller communities.

■ 1.6   END-USER PRESSURES

From the broadest viewpoint, that of society as a whole, one can make the argument that extending fiber to the premises is not a mere luxury but almost a necessity. Nobody needs to be reminded of the present stagnation in the telecommunications business in which significant excess capacity in the interoffice facility backbone remains unused, in spite of constant growth in number of available WDM channels and constantly dropping WDM terminal costs. But there is also stagnation at the premises end, evident in the slow introduction of HDTV transmission (relative to the wide availability of HDTV rentals, players, and even camcorders). An even more important stagnation is that occurring in the computer part of the overall information industry, as evidenced by the fact that the time spans between new generations of both hardware and software are growing ever longer, and also by the fact that few of these innovations are communication based. Note that the World Wide Web, the last truly significant communication-based innovation, is more than 15 years old.

A computer or TV terminal is much more than a gigabit per second window into the contents of the hard drive; it is (or can be) a gigabit per second window into the entire information world. When one sees backbone telecommunications capacity going to waste, while the computer industry's innovations, such as they are, are almost completely intramachine or low-speed intermachine, one begins to sense the economic potential and societal benefit in opening up the intervening bandwidth bottleneck of Figure 1.1.

■ 1.7   SPECIFIC END-USER APPLICATION NEEDS

Let us examine more closely the claim that per-user needs for bandwidth are sure to continue to grow. After all, ever since Shannon and Pierce analyzed the bandwidth of the human perception mechanism in the 1940s, one hears the recurrent claim that, psychophysically, people simply cannot absorb much more information per second than they already get with telecommunications. This is a claim that has been overturned so many times that nobody really believes it any longer.

A semilogarithmic plot of premises and long-haul bandwidth capability compared to access bandwidth is shown in Figure 1.5. Like all such quantifications, the bit rates shown are the intermittently required bit rates—it is not implied that each user needs the full bit rate all the time without sharing based on intermittency of need. It is seen that the four orders of magnitude bottleneck of Figure 1.1 has been with us for a long time and is not getting any better. New applications seem constantly to emerge that require the mitigation of this bottleneck in order for society to progress.

What are these driving forces today? Many of today's broadband applications are being driven by Internet users, others by broadcast television, and still others by Hollywood. They can all be broken down into two classes:

- Medium-size file transfers absolutely requiring low latency, for example, broadcast television, interactive and conferencing video, security video monitoring, interactive games, telemedicine, and telecommuting.

- Transfer of files whose latency is not so important but for which long transfer times are annoying: video (movies) on demand, video and still-image email attachments, backup of files, program sharing, and downloading, for example, of entire books.

There is not only a bandwidth demand from existing and well-understood applications but also a huge poorly understood opportunity to develop new applications and services. Let us discuss the existing pressures first.

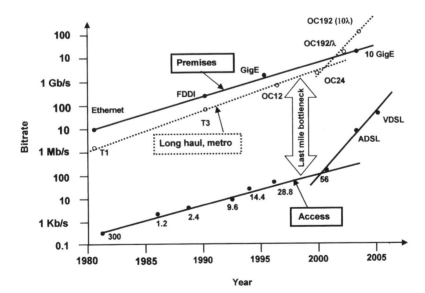

FIGURE 1.5 Historic view of the persistent bandwidth bottleneck of Figure 1.1. Updates from Prof. L. Kleinrock.

Perhaps the most significant well-defined driver for FTTH today is high-definition television. While conventional television (**standard-definition TV, SDTV**) requires 20 Mb/s uncompressed and 4 Mb/s per channel when compressed with MPEG-2, HDTV requires 120 MB/s uncompressed and about 15 to 19 Mb/s when compressed with MPEG-2 [Poynton]. Multichannel HDTV is beyond the reach of DSL today, except for the shortest distances betweeen head end and subscriber, while cable modems offer only a slightly better solution. Current HDTV offerings, for example, by satellite broadcast, are limited to a small integer number of channels, whereas the public has gotten used to a choice among hundreds of satellite or cable SDTV channels. One artifice that is being studied by cable providers and telcos alike is not to try to send many HDTV channels to subscribers, but instead only one that has been preselected by the subscriber (**video on demand, VoD**), a temporary solution at best.

Equally important, and already an issue, is the growing need for fast response and download/upload times between desktop or laptop computers or between them and central facilities such as search engines. When the data objects to be exchanged reach significant size, the response time limitations are due not to hand-shaking, application execution, or file access delays but to the time taken to transmit the large data object to the user. This is a problem today only for the largest such objects as large documents and images but is already getting much worse as object size grows to the level of entire books or even entire movies.

For example, online movie rental on demand and the exchange of home videos are already appearing as real modes of system usage. Today such applications suffer from intense user dissatisfactions with a delivery time of several hours to download a full movie, even at cable modem speed and distance. A trip across town to Blockbuster is quicker. Similarly, mailing a CD-ROM containing many pictures or a roll of film is usually a better solution than emailing it. We have all suffered through watching video clips of postage stamp size on our PCs when what we really want is full-screen images.

With the growth of telecommuting and off-site use of home office resources, it is often necessary for entire "projects" to be moved quickly to some unexpected new location in order for the person's work to continue.

Another current application is remote disk backup at centralized secure servers, where the remote central facility is mapped as a local drive. At the PC level, this has the potential of displacing today's large sales of Zip drives, writeable CDs and DVDs, and the like. At the commercial data processing (DP) level, this remote backup application has been around for a long time and was the initial motivator for the earliest DWDM systems. Now yesterday's point-to-point connection of a data center to a remote backup facility has evolved into the **storage area network (SAN)**, a more topologically complex structure aimed at even better data availability, high speed, and low-latency performance. The potential of this application is constrained today by the excessive time required to communicate large files, to say nothing of the contents of the entire PC or data center.

Response time plays a key role in human productivity in using computers, in a way that is not always appreciated. Some years ago a classic but little-noted psychophysical study [Thadhani] showed the importance of response time for many kinds of multistage processing jobs. When the response time of each step goes below several hundred milliseconds, in the user's mind the individual stages begin to merge into one simple stage, and a surprisingly large throughput increase occurs. Thus, one reason today's last-mile bandwidth bottleneck is such a constraint is the increased size of the data objects that people use in multistep communication-based information processing. The pressure is to replace email with voice mail or video messages, thumbnail images with full-screen images, and static images with video. And yet to keep the productivity of the single-step continuum mode of multistep job handling, we need to be able to communicate all these objects with the same response time we have learned to live with for text commands or mouse clicks for local execution.

As time goes on, the increasingly peer-to-peer nature of Web traffic will become more evident. The growth of symmetrical traffic loads, compared to the traditional asymmetrical case, has been widely noted and has been attributed to the rise of Web-based traffic and the proliferation of blogging and other modes in which the center is in a PC, not at some large information service provider. The example most often cited is music file sharing provided by Napster and its successors. Each of today's standalone ways of handling numbers, words, correspondence, images, and games can be seen to represent a special case of a larger universe of peer communication-based versions of the same thing. For example, sociologists have frequently commented on the antisocial nature of children's battles with aliens inside their own machine versus having your aliens battle someone else's aliens. Today, multiparty games have the needed resolution but not the needed rapid response time.

Another point concerns videoconferencing and group collaborative processing. The long-standing vision of being more able to "exchange communication for transportation" has taken on a new urgency since the events of 9/11. The global nature of the workforce, with many people now living and working in one state or country for employers in another, today leads to separations of families and other situations that adequate bandwidth would alleviate. As for conferencing, the insufficiencies of the available video conferencing and group collaboration technology send a large number of people onto airplanes to attend meetings of fairly simple structure for which today there is no adequate electronic substitute that provides the needed full interpersonal communication between multiple participants, complete with eye contact and other cues. The missing link is communication bandwidth since promising high-resolution display and stereo hi-fi audio technologies are well in hand.

Then there are the unpredictable applications. While history shows that it is usually some unpredicted application that later becomes the most important one, it also shows that it is the predictable ones, like those just listed, that pay for the initial transition. For example, while email turned out unexpectedly to be the most important use of the Internet's parent, the ARPAnet, the promised resource sharing of a few very costly specialized computers was enough to justify ARPAnet's introduction economically.

Some pundits think that breaking the bandwidth bottleneck of Figures 1.1 and 1.5 with fiber will be so significant as to constitute a "Second Coming" of the PC, the first one having been the 1980s revolution that empowered localized individual use of computer power. Closing the last-mile bottleneck seems highly likely to ignite a takeoff of communication-based multi-party and collective information processing. Many of these applications will be new ones that are difficult to envision today. One may count on the inventiveness of Silicon Valley and Redmond, Washington, to think of new bandwidth-hungry things that we shall subsequently wonder how we ever lived without.

■ 1.8 THE DIGITAL DIVIDE

The term *digital divide* has been used to describe the effect of ever more sophisticated information technology on making the poor poorer and the rich richer. It has to do not just with economic ability or inability to buy the technology, but also the availability or unavailability of help in understanding the technology well enough to use it. The replacement of the DOS **C:** prompt with graphical user interfaces somewhat reversed the trend toward decreasing user friendliness with PCs, a useful reminder that sometimes more technological complication inside can be used to deliver more simplicity looking from the outside. (As another example of this deliberate exchange of internal complexity for external simplicity, consider the automatic transmission as compared to the manual version.)

Reasoning by nothing more than analogy, it may not be naive to hope that more bandwidth will mean more ease of use simply by removing some of the complications imposed by transmission media limitations. Anecdotal evidence supports this conjecture. For example, business writer Charles Ferguson [Ferguson] reports visiting a Brazilian city slum and seeing large numbers of children queuing up to use PCs, but getting minimal service because 10 PCs had to share a single phone line.

It is to be hoped that public policy pressures that emphasize the removal of bandwidth limitations to the poorer levels of society, not just the more affluent, will have the same strong empowering effect that bringing Windows-based PCs and Macs into disadvantaged schools and homes has had.

The Universal Service Fund, to be described in Section 5.2 has been a significant step in serving the rural American populace.

Based on the notion that bringing very large access bandwidths to large segments of the population is not only good business but good citizenship as well, top executives of many of the largest North American information companies have formed an interesting consortium called TechNet [Technet]. Members include HP, Cisco, Intel, 3Com, Microsoft, and others. One of TechNet's objectives is to "encourage broadband deployment to underserved communities and businesses through investment incentives . . ." They call for "affordable 100 megabit per second broadband connection to 100 million homes and small businesses by the year 2010."

As we shall see in Chapter 5, the initial deployments of FTTH that have taken place in the United States have been to small, less advantaged communities, and the deployment has been by small companies. This "bottom-up" evolution, by the way, was the way cable TV started out. The large U.S. telecomm carriers, however, are employing a "top-down" or "cherry-picking" strategy of serving affluent suburbs first. The plan seems to be that eventually FTTH service from the large carriers will "trickle down" to the less affluent areas.

■ 1.9 COST IMPROVEMENTS

The replacement of last-mile copper by fiber has important positive economic consequences for both provider and end user, not only for first cost but especially for lifetime costs. Consider the typical PON or home run system of Figure 1.3(b).

In most optoelectronic systems, the cost of the optics dominates the cost of the electronics, and FTTH systems are no exception. First cost plays a dominant role in the last mile to a much greater extent than with the more traditional metro and long-haul situations, which are supported by aggregated revenues from many thousands of subscribers. The expensive optical items (which we shall discuss in Chapter 3) include the lasers, especially the highly linear 1550-nm lasers required at the central office for frequency-division multiplex (FDM) analog TV service, the optical amplifiers, the lower powered lasers at the subscribers, and the splitters and WDMs associated with the coarse WDM basis of most system architectures. Photodetectors and their associated electronic amplifiers are less of an economic factor. The fiber cable itself is cheaper on a per-strand basis than copper.

The optoelectronic equipment cost is a significant enough fraction of the total cost in the last mile that the future of FTTH is going to depend significantly on further cost reduction in the optics. For example, one 2002 estimate [Ponder] cited 15% of the cost as lying in central office optoelectronics, 40% in the distribution network and its installation, and 45% in the customer premises optoelectronics and its installation. The same source broke down the 40% for the infrastructure as follows: 53% in construction, 10% in engineering, 20% in couplers and splitters, 9% in splice closures, and only 8% in the cost of the fiber itself.

A lot of optoelectronic cost reduction is happening. The same kind of creativity that has been applied by the optical networking technology fraternity for the last decade to increasing functionality is now being applied to the problem of cost. For example, passive splitters, which cost over $100 per port a year or two ago are now available at $13 per port in quantity.

While clever things are happening in the optical component world, equally inventive things are happening in the civil engineering of fiber installation, both aerial and underground. These advances are discussed at length in Chapter 4. For example, the installation cost for underground FTTH facilities dropped to about $5 per foot by 2004, averaged over a combination of open trenching, plowing, and directional drilling as circumstances require. Aerial construction is as low as $1.50 to $2.50 per foot [Render].

Thus, the costs, particularly the lifetime costs, of the all-glass solution have dropped to levels that are comparable to or less than those of any of the copper-based solutions, for one thing because the latter include along the right-of-way hot, finite-lifetime electronics with localized backup power sources placed at intervals along the right-of-way. Much of the electronics cost is for very clever coding and equalization schemes, and for data compression and decompression, unnecessary with fiber because of its huge offered bandwidth.

Probably the biggest cost savings in introducing FTTH have to do not with technology but with network management and business management. The benefits to the service providers of **service integration**, in which all services are lumped into at most two businesses, wireless and fiber, are now seen as one of the more important reasons for going to triple play over fiber. A recent authoritative joint study of the **capital expenses (CAPEX)** and **operating expenses (OPEX)** of Verizon, SBC, and Bell South by Bernstein Research and Telcordia [Bernstein] reached several interesting conclusions:

- Ever-decreasing cost to deploy should drop to about $1400 per customer averaged over the years 2004–2008.

- Extending FTTH to 50% of these three carriers' residential customers and 80% of their business customers in suburban areas of the top 20 U.S. states (costing the carriers $45 to $50 billion) would improve their revenues by $9.5 billion by the end of the deployment, which would take 5 to 7 years.

- Annual savings in operating expenses would reach $5.5 billion by 2008 and $7.8 billion by 2010.

- These OPEX savings would be gained by such things as a 100% expense reduction at the central office (CO), and 30 to 70% reduction in customer service and network maintenance expenses.

In other words, the study recommends that these incumbent carriers undergo a temporary CAPEX bathtub in exchange for a permanent OPEX and revenue improvement.

Seemingly trivial things contribute heavily to the cost of running a traditional copper-based circuit-switched POTS-dominated telecommunications enterprise. Many of these are reduced or eliminated by substituting a frozen but highly versatile glass infrastructure between CO and every subscriber. For example, the costs for technician dispatches and low-value customer contacts for customer churn and for repair are greatly reduced with FTTH. Churn, defined as an economically harmful amount of reconfiguration, is a big factor with today's telco plant, partly because there are several manual cross-connection points between CO and subscriber phone jack, and making any change in these panels requires at least one technician and one truck roll for each.

■ 1.10 NEEDS OF THE SUPPLIER INDUSTRIES

The evolution from last-mile copper to last-mile fiber is already presenting many new needed business opportunities for many different sectors of the economy. As technology, application, economic, and regulatory forces continue to drive the evolution to FTTH, there is a hierarchy of solution providers who must be considered in forming a total picture of the way in which this set of developments serves various needs. The next three chapters of this book are ordered in such a way as to treat the technical aspects of these elements in sequence.

- System architects and writers of code (Chapter 2)
- Base technology manufacturers: fiber, low-cost uncooled lasers, and the like (Chapter 3)
- Box manufacturers: central office OLTs or OLT blades, subscriber ONUs, set-top boxes, and so forth (Chapter 4)

- Installation technology developers: trenching, fusion splicing, test equipment, and so on (Chapter 4)
- Installers: turnkey trenchers and hangers of fiber (Chapter 4)
- System purveyors: contractors and subcontractors (Chapter 5)
- System owners (Chapter 5):
  - Large ILECs, in the United States these being the regional Bell operating companies—Verizon, Southwestern Bell Company, Bell South, and most of Qwest
  - Electrical utilities, state or locally owned, numbering about 2000 in the United States, with 40 million customers in communities of size typically less than 10,000
  - Small ILECs, numbering about 1400 in the United States
  - Multiple system operators: cable TV system owners
  - Overbuilders (second system owners in an area where an incumbent local exchange carrier or cable company preexists)
  - Content providers (broadcast networks, news organizations, entertainment providers, etc.)

The economics of each of these sectors is being affected positively by the opening up of the last-mile bandwidth bottleneck. For the limited purposes of this book, we shall not try to predict the effect of FTTH on the last one, the content providers.

■ 1.11 NEEDS OF THE TELECOMM SERVICE PROVIDERS

In the acronym ILEC, the modifier "local exchange" is rapidly becoming less appropriate as the large ILECs buy up failing long-haul **interexchange carriers (IXCs)** and expand into other businesses. Nevertheless, in the absence of newer terminology, we shall refer to Verizon, Bell South, SBC, and Qwest as the large ILECs, and MCI, AT&T, and Sprint as IXCs. As of late 2005, MCI and AT&T have been absorbed by competing ILECs. It is the local exchange or *access* part of these companies' businesses that is of interest here.

Like that of any other industry that consumes the public's resources, regulation dominates telecommunications and can either be an inhibitor or an enabler. The common carriers of the world have spent heavily to make sure that regulation becomes an enabler of their welfare, but this has often acted as an inhibitor of the public welfare.

The heavy hand of regulation is particularly apparent when the resource in question is not easily sharable between owners, for example, radio spectrum, synchronous satellite orbital positions, and highway and airspace rights-of-way. In such cases of unsharability of a resource that was already there, a **natural monopoly** became a permitted solution. This concept is being steadily and significantly eroded in some interesting cases. For example, since the use of spread-spectrum techniques allows a number of users to occupy the same radio frequency spectrum with negligible mutual interference, as with WiFi, Bluetooth, or spread-spectrum cellular telephony, RF spectrum is no longer always intrinsically subject to natural monopoly rules. For decades, wireline telecommunication was considered a natural monopoly. At one time it was even difficult to get power and communication companies to share the same utility poles. For years now, it has been standard practice that poles can carry power, coax, telco copper, fiber, alarm system wiring, and so forth.

In some cases governments now seem to feel that they went too far in undoing perceived natural monopolies. For example, until recently, starting with the Telecom Act of 1996, the U.S. federal regulators decreed that the ILECs must provide unbundling—also known as UNE-P (unbundled network elements—platform). An unbundled element was one that the regulated carrier was obliged to make partly available to a competitively disadvantaged competitor. The word "platform" was taken to mean not only line sharing (sharing bandwidth in the copper) but also sharing of space at the CO.

The Federal Communication Commission's (FCC's) 2004 Triennial Review Order said that:

- For FTTH facilities to new homes (the **greenfield** situation), no unbundling to accommodate other carriers was required.

- For FTTH facilities that were **overbuilt** (added) to existing copper facilities, as long as the copper was removed, only the voice component had to be unbundled. If not, all services were subject to the unbundling requirement that had existed since 1996.

The root of the regulatory impairment to the carriers in the United States can be traced to the decades-old fact that the telcos are classed as Title I ("telecomm services") providers, whereas cable providers are Title II ("information services") providers. This has given the cable providers almost complete freedom from regulation. In mid-2005, the U.S. Supreme Court declined to erase this distinction, a large victory for the cable companies.

Thus, the large U.S. ILECs, facing a choice between conducting "business as usual" and thereby steadily losing their terrestrial business to cable (as discussed in Section 1.3 and Fig. 1.4) or reinventing themselves, seem to have chosen the latter. This has meant becoming more technically aggressive and (as mentioned in Section 1.9) streamlining business procedures and likely adopting the strategy of a short-term CAPEX bathtub to enable a long-term revenue and OPEX improvement. Regulatory relief has helped.

As analyzed in [Bernstein], this reinvention process, or "process reengineering," some of whose benefits we described in Section 1.9, faces several barriers:

- The personnel downsizing enabled by the less labor-intensive FTTH technology will be more difficult than downsizing has been in other less heavily unionized industries. While FTTH has its human upside for users, this is not necessarily so for members of the telco workforce and their families.

- Regulation. Here, the carriers' tradition of aggressive intervention in the regulatory process using armies of lawyers and lobbyists seems to have either been addressed to yesterday's issues or has backfired as far as the regulators have aggressively defended their prerogatives. It seems to this writer that, for various reasons, including incomplete technical literacy, the regulators often lag the carriers, who in turn often lag the supplier industry in getting their minds around the content and implications of new technology evolutions. (The research arms of the telephone companies have led the world in Nobel prizes but have famously been no match for Silicon Valley-style entrepreneurship in delivering to society the fruits of invention and innovation).

- Among the other legacy burdens from the era of heavy common carrier regulation is the lingering obligation for the ILECs to contribute to the USF and to provide unprofitable **lifeline phone service** to the aged and underprivileged.

- Local and federal tax surcharges remain on the ILECs services that do not exist on cable service, even when the latter included POTS.

- A conservative management mind-set that must compete with the more Silicon Valley-like or Hollywood-like management of the cable companies.

Other countries face other regulatory problems. For example, in Japan the dominant landline carrier, NTT, is legally prohibited from TV program distribution.

In spite of the presence of these disincentives for new expensive undertakings by carriers, particularly FTTH, it appears that these problems are in the process of being solved, as evidenced by the deployments that we shall discuss in Chapter 5.

■ 1.12 DEFICIENCIES OF THE LEGACY SOLUTIONS—DSL, CABLE, AND WIRELESS

Let us examine quantitatively the legacy last-mile broadband technologies in light of the needs of the end users and those of the suppliers and the facilities providers that we have been discussing. Figure 1.6 shows the orderly historic progression of fixed wireline communication from the earliest star-configured solutions of the telcos and cable providers to the present wave of fiber-based PONs.

In the prefiber days, telco COs were interconnected by coax and microwave, while cable head ends were fed by microwave or satellite. The first step in "fiberizing" the entire system was either **hybrid fiber–coax (HFC)**, which has been widely deployed by the cable industry, or various forms of DSL, its telco twisted-pair competitor. As Figure 1.6 shows, both cable providers and telcos use fiber as a feeder for remote equipments that feed individual subscribers. These equipments are *nodes* in the cable case and *subscriber line multiplexors* of various flavors in the telco case.

The term **fiber to the node (FTTN)** is often used for what is essentially the HFC topology. Ultimately in the evolution, the HFC node becomes a very high speed digital subscriber line (VDSL) optical network unit (ONU) [DSL].

Today, HFC is still the most widely deployed broadband solution of the cable companies, while **asymmetric digital subscriber line (ADSL)** dominates the telcos' installed base. Because of the inferior intrinsic capacity of 24-gauge legacy copper going back to Alexander Graham Bell, compared to that of coax, basically a post–World War II technology, DSL has been competitively successful mostly for short distances and low bit rates. Some figures on the current-installed

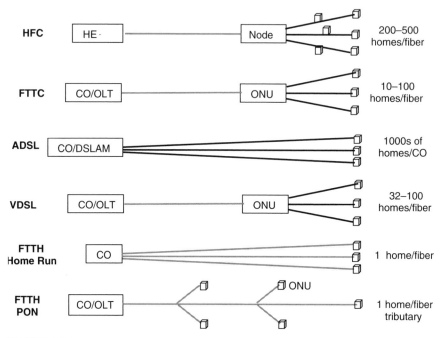

FIGURE 1.6 Evolution of terrestrial broadband from one fiber appearance per 500 homes to one for every home. Black lines indicate copper, gray lines indicate fiber.

base of fiber last-mile solutions in various countries will be given in Chapter 5, but suffice it to say here that at present in North America, where access distances are typically larger than in many other countries, more that twice as much digital broadband cable service has been installed as DSL.

The distance and bit rate capabilities of the various forms of DSL and cable modems are summarized in Figure 1.7. One should carefully note that the throughput numbers for the bottom five entries in Figure 1.7 are *aggregated* totals across all the subscribers attached to one tree, while the top three entries are per-subscriber numbers, provided the subscriber connects directly to the CO.

Digital subscriber line (DSL) has gone through many generations [DSL] of ever more sophisticated signal processing to squeeze more bit rate and distance from the traditional 24- or 26-gauge twisted-pair subscriber loops. Figure 1.8 gives more detail on the bit rate–distance trade-off for various DSL dialects, and it is seen that these distances are quite modest at interesting bit rates, considering the emerging requirements we have been discussing, for example, HDTV. The figure also shows the capacity of today's cable modems and of one form of FTTH, namely the **broad band passive optical network (BPON)**.

The distances given are those from the DSL node (**DSL access multiplexor– DSLAM**) to the subscriber. The DSLAM usually lives at the central office. The good news about the figures given is that the bit rates are unshared bit rates. The bad news is that if one tries to extend the distance by interposing a remote

Service	Medium	Mb/s Down	Mb/s Up	Max. Dist at Full Rate (km)	Standard
ADSL	Twisted pair	8	0.64	2.4	ITU G.992 series
VDSL	Twisted pair	40	6.4	0.4	ANSI T1E1, etc.
ADSL2+	Twisted pair	16	1.0	1.5	ITU G.992.5
HFC	Coax	57	9.2	25	DOCSIS 2.0
BPON	Fiber	155 or 622	155	20	ITU G.983 series
GPON	Fiber	1244/2488	155–2400	20	ITU G.984 series
EPON	Fiber	1250	1250	20	IEEE 802.3ah

FIGURE 1.7 Comparison of various forms of DSL, cable modems, and passive optical networks. DSL bit rates are per subscriber and HFC and PON bit rates are shared among subscribers.

DSLAM to get nearer the subscriber, then the **backhaul** from remote DSLAM to the CO becomes a shared facility, so often very little has been gained. (The backhaul is typically T-carrier or Ethernet, although there has been discussion of going to Gigabit Ethernet.) Thus, unless and until DSLAMs are remoted from the CO over multi-Gb/s services that can handle the aggregated traffic, the inefficient line sharing that is avoided by the star connection from DSLAM to customer is simply moved upstream toward the CO.

From the beginning, the intent of DSL has been to provide a useful data path that can operate concurrently with the customer's POTS. The latter uses the lowest 3 kHz of the available spectrum. The DSL evolution has culminated (at present writing) in **ADSL2+**, which was approved by the International Telecommunications Union in 2003. So far, the penetration of ADSL2+ has been quite small in the United States. By far the most widely used DSL alternative is ADSL, which can be seen from Figure 1.8 to have a much longer reach but less capacity than ADSL2+. Obviously, the principal difficulty in using any form of DSL to compete with coax (much less fiber) has been the very limited reach available from such an unpromising medium as unconditioned copper twisted pair. Somewhat more distance might have been squeezed out of the transmission medium by retrofitting each line with conditioning in the form of loading coils along their length (as has been done for 1.5-Mb/s T-carrier transmission for businesses), but this option would have been prohibitively expensive for service from the DSLAM to single residences.

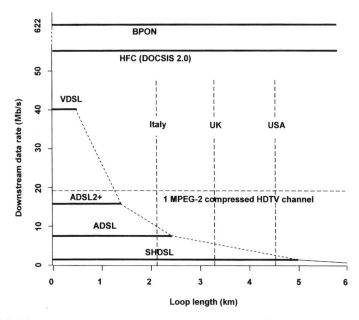

FIGURE 1.8 Maximum data rate vs. reach for BPON, DOCSIS-2 cable, and various forms of DSL. Vertical dashed lines show the distances covering 85% of potential subscribers for three countries (from Fig. 1.2).

The emerging requirement of more capacity symmetry, driven by many factors including the proliferation of many Internet and Web hubs in residences and small businesses, has constituted a significant technical challenge for both DSL and cable. Upstream rates still remain quite low. As we shall see in Chapter 5, fiber to the curb (FTTC), serving several tens of subscribers per fiber end, is being implemented by some U.S. carriers in preference to FTTH, and VDSL or ADSL2+ is expected to prove a useful vehicle for running the last few hundred feet in copper.

Perhaps the inadequacies of DSL can best be summarized by noting from Figure 1.8 that yes, high bit rates can be carried by copper twisted pairs using sophisticated signal conditioning and far-end processing; but no, this will not address the distance needs shown in Figure 1.2 and the bandwidth growth needs shown in Figure 1.5.

Hybrid fiber–coax, serving several hundred homes per fiber end, using bidirectional copper coax between "node" and subscriber, is the favorite solution of the cable providers. The topology of the network is a P2MP tree, as shown in Figure 1.3, and this produces many challenges not faced with the DSL P2P (point-to-point) geometry of DSL.

Hybrid fiber–coax has evolved rapidly. The DOCSIS 2.0 standard [DOCSIS] dates from 2001; yet premises cable modems based on this standard are about the only thing being installed for digital cable service today. **DOCSIS** stands for "data over cable service interface specification." It is quite obvious from examining Figure 1.8 that the rapid installation of such cable modems in suburban service areas that are unreachable with any form of DSL that has interesting capacities presents a significant competitive threat to the telcos.

Both DSL and cable modems provide a vivid example of how throwing every conceivable form of signal processing at a medium with many pathologies can make a usable silk purse out of a particularly unpromising sow's ear. Among the many phenomena afflicting the copper plant are certainly its limited usable bandwidth (roughly 0.75 GHz for coax) and the nonflatness of its attenuation spectrum, but also pickup of impulse noise and hum, and the fact that such interferences accumulate going upstream in any P2MP topology.

The coax-based HFC architecture bears a very close resemblance to that of the optical PON. This is true in its P2MP geometry and also in many of the protocol details. However, there is a very important distinction. As the reader will see in Chapter 3, sophisticated signal processing plays essentially no role in a PON because of the fiber medium's low attenuation, wide bandwidth, and lack of any forms of noise other than white Gaussian noise in the receiver front end. Signal processing, however, is key to HFC performance. The only real complication is managing the bandwidth sharing in the upstream direction, an issue that PONs share with HFC and for which HFC architects provided the template [Ransom].

■ 1.13 FUTURE-PROOF NATURE OF THE FIBER LAST MILE

There are several reasons for claiming that the fiber last mile, once installed, is unlikely to require any reengineering for decades. This places it in a different category from DSL and cable systems in which hot, protocol-sensitive electronics intervene between head end and subscriber.

The first aspect of this "future-proofness" is fiber's enormous available and scalable room for growth in bandwidth (which we shall quantify in the next chapter) as well as its freedom from crosstalk and from tappability by intruders.

Also, it is clear that fiber's low attenuation translates into low-power transmitters and relatively insensitive receivers, but what may not be so clear is that it also translates into much more forgiving design rules for the installation than those available for DSL or cable. For these legacy media, great care must be taken about segment length, selection of cable type, variable attenuator and equalizer settings, and so forth. (The author's cable modem installation required segments of three different coax types just between street and modem, three truck rolls, and an hour of system adjustment of the equipment in the overhead enclosure.) Experiments are under way to do some of the provisioning from central office or head end.

The passive nature of the PON medium and the fact that the electronics is only at the ends means that provisioning, reprovisioning, and repair are accomplished much more quickly than with systems embodying electronics along the right-of-way. Already, the carriers who are installing early versions of the available PON architectures (specifically BPON, as discussed in the next chapter) are saying that when the time comes to upgrade the system to ATM-based gigabit passive optical networks

(GPONs) or Ethernet-based passive optical networks (EPONs), they will do this mostly by remote software download to all the ONUs from the central OLT.

As with dense DWDM optical networks, the glass is transparent with respect to legacy or future traffic types. Thus, as new bit rates, formats, higher protocol levels, and services are introduced, nothing needs to change in the **outside plant**, the portion of the system lying between head end and subscriber—only optoelectronic terminal equipment. Historically, it has always been the case that the need to support protocol variegation becomes more acute the farther one progresses outward from the telco backbone, which uses either legacy SONET/SDH or is going to Internet Protocol with Multiprotocol Label Switching (IP/MPLS) in the near future. One may expect more protocol churn in the access. Fiber is the ideal medium to make sure that the hit is only at the premises and the CO.

It was the protocol insensitivity of both telco copper and of wireless that allowed the proliferation of such an imaginative repertoire of services as we have seen. The same may be said of the Internet, because the IP protocol is so simple that it was easy to build a variety of higher protocol layers on top of it. So much so that interest in the art of formal protocol conversion [Pconv] has more or less died out completely.

■ 1.14 WHY BRINGING FIBER ONLY TO THE CURB IS INSUFFICIENT

As outlined in Section 1.12, today's ADSL2+, VDSL, and HFC are capable of several tens of megabits per second total throughput. However, since for HFC this throughput is spread across the population of attached subscribers, who may number in the hundreds, it is clear that these capacity figures are already being overtaken by the applications enumerated in Sections 1.6 and 1.7.

Passive optical networks, on the other hand, even though similarly based on a shared medium, possess at least 10 times the capacity at usable subscriber distances and require no intervening electronics, a reliability and cost Achilles' heel of DSL and cable.

There has been a steady decrease in the first cost of FTTH compared to the copper solutions, to the point where fiber costs slightly less. As we shall see, this is not only due to decreasing material costs but also to much more manageable ways of deploying, splicing, and terminating multifiber cable. The really significant cost advantage comes when lifetime costs must be included. At this point, the absence of electronics along the right-of-way, whether near to the CO or near to the subscriber, ceases to be just an aesthetic argument and becomes a very practical one. It means that no matter how short the copper portion there must be some powered electronic elements. As for powering, it seems more of an onus to run power to a remote fiber-fed DSL electronic node than to ask the FTTH subscriber to use his wallplug power with battery backup for his electrooptical ONU.

A detailed comparative study of all cost elements of DSL, FTTH, and FTTC, confirming the business advantages to the providers of going for FTTH is given in the Bernstein/Telcordia report [Bernstein].

1.15 THE WIRELESS "ALTERNATIVE"

Wireless access and fiber access appear to be mutually complementary technologies. Fiber serves specific fixed locations with enormous capacity and reach, while wireless can go everywhere but with very limited capacity, variable and uncertain quality of service, and high density of base stations. Some of the numbers make the point [Varma, Desurvire]. Bluetooth spans several meters. Wifi 802.11 wireless local access networks (LANs) support up to 54 Mb/s up to 30 m and lower bit rates up to 300 m. Third-generation cellular telephony such as the evolving Global System for Mobile Communication (GSM) and General Packet Radio Service (GPRS) and others have as their target 2 Mb/s over a wide range of distances. Even if the bit rates of such technologies are increased, for example, by decreasing the radius of radio coverage, there is still the problem of providing a terrestrial backhaul connection to a CO for each cell. With FTTH, the backhaul stays where it has been—up the hierarchy from the CO.

One striking thing about wireless has been the imagination that has built such a useful portfolio of services on such an unpromising medium: telephony, fax, text, voice mailboxes, text messaging, stock quotes, global positioning system (GPS) position location, rudimentary video, even speech recognition for hands-free usage. There is no reason to think that opening up the last mile of fixed telecommunications using a protocol-insensitive medium will have an impact any less extensive and exciting.

A second striking thing about wireless, and one that is not widely appreciated, is that it is beginning to top out. Whereas the industry grew 23% in 2001, by 2003 the figure had dropped to 8% and is predicted to fall to 4% by 2008 [Bernstein].

One last fixed-path last-mile technology should be mentioned, the use of RF transmission to residences and offices over existing power lines. Even though this form of "last-mile" transmission is cheap, requiring no new paths and allowing access simply via the existing wall plug, the data rates are quite low, 2 to 6 Mb/s at present [Mollenauer] [IBM].

1.16 THE POSITION OF THE SKEPTICS

Not everyone feels completely enthusiastic about replacing all that copper with fiber. As mentioned, some feel that bringing fiber only to the curb or relying on wireless will be a solution that will endure for a long time. The existing capital investment in copper is enormous. Performance of DSL and cable has been improving steadily, so why do we need anything else? Likewise for wireless. The emergence of applications requiring multiple tens of megabits per second per user is only gradual, so let us wait and see. These PONs don't really address the future—just look at the sharing of capacity and the failure to standardize analog TV, which will be with us almost indefinitely. FTTH does not provide the central powering for "lifeline" continuous service that copper provides and is therefore unacceptable. The imminence of fiber to the home has been repeatedly claimed since the 1970s, so what is new about the present wave of claims and pilot studies whose principal purpose seems to be to placate the regulators, investors, and futurists? The fact that desktop and laptop computers are so powerful argues that we need less access to remote data objects and execution capability, not more. The economy was harmed, not helped, by overly enthusiastic predictions a decade ago of the magical consequences of all-optical networks and photonic switching centers. So isn't this just the same sort of wishful thinking at work? Needless to say, I don't believe a word of this, and my point-by-point reasons, if not clear already from this chapter, should become clearer as this volume proceeds.

Like Mount Everest, the principal thing that can be said for DSL and cable is that "they are there." But when it comes to spending large new sums in the presence of available FTTH technology and opportunities for new revenues based on new services to the public, this argument hardly suffices.

REFERENCES

[Bernstein] *Fiber: Revolutionizing the Bells' Telecom Networks*, Joint Sanford C. Bernstein/ Telcordia Technologies, New York, May, 2004.

[Desurvire] E. Desurvire, *Global Telecommunications—Signaling Principles, Network Protocols and Wireless Systems*, Wiley Interscience, Hoboken, NJ, 2004, Chapter 4.

[DOCSIS] The Cablelabs website, www.cablemodem.com, contains detailed information on DOCSIS 1.1 and 2.0.

[DSL] www.dslforum.org.

[Ferguson] C. H. Ferguson, *The Broadband Problem—Anatomy of a Market Failure and a Policy Dilemma*, Brookings Institution Press, Washington, 2004.

[Green] P. E. Green, Jr., *Fiber Optic Networks*, Prentice Hall, Englewood Cliffs, NJ, 1991.

[IBM] IBM-Center Point joint venture, NY Times, July 11, 2005.

[Mickelson] H. Mickelson, E. Sundberg, P. Strömgren, and Y. Fujimoto, Loop lengths and architecture, *IEEE EFM Conference*, Raleigh, Jan. 14–16, 2002.

[Mollenauer] J. F. Mollenauer, Broadband over Power Lines, presentation at OFC/NFOEC Conference, Anaheim, CA, March 9, 2005.

[Pconv] P. E. Green, Jr., Protocol Conversion, *IEEE Trans. Commun.*, Vol. COM-34, No. 3, March, 1986, pp. 257–268.

[Ponder] D. Ponder, FTTH Spells Success for an Independent Telco, FTTH Conference, New Orleans, October, 2002.

[Poynton] C. Poynton, *Digital Video and HDTV*, Morgan Kaufmann, San Francisco, 2003.

[Ramaswami] R. Ramaswami and K. Sivarajan, Optical Networks, 2nd ed., Morgan Kaufmann, San Francisco, 2002.

[Ransom] A. Azzam and N. Ransom, *Broadband Access Technologies,* McGraw-Hill, New York, 1999.

[Render] *Fiber to the Home, The Third Network—2004/2005*, Render, Vanderslice and Associates, Tulsa, December, 2004.

[Thadhani] A. J. Thadhani, Factors Affecting Programmer Productivity During Application Development, *IBM Sys. J.*, Vol. 34, No. 1, 1984, pp. 19–35.

[Technet] www.technet.org.

[Varma] V. K. Varma, K. D. Wong, K.-C. Chua, and F. Paint, Special Section on Integration of 3G Wireless and Wireless LANs, *IEEE Commun.*, Vol. 41, No. 11, November 2003.

a VOCABULARY QUIZ

Discuss not only what these terms abbreviate but also what they mean.

Access	FTTN	Metro	PON
ADSL	FTTP	MPEG-2,4	POTS
APS	Greenfield	MSO	SAN
Backhaul	HDTV	Natural	SDTV
CAPEX	Homerun	monopoly	Service
Churn	connection	NID	integration
Dense WDM	HFC	ODN	Star
Digital divide	ILEC	OLT	connection
Direct fiber	IOF	ONT	Triple play
connection	IXC	ONU	VDSL
DOCSIS	Last-mile	OPEX	VoD
DSL	bottleneck	Outside plant	VoIP
DSLAM	Lifeline phone	Overbuild	Wavelength
DWDM	service	P2P	division
FTTH	Long-haul	P2MP	Wireline

CHAPTER 2

Architectures and Standards

▪ 2.1 INTRODUCTION

The FTTH installations that are going in today are either of the home run (P2P) or splitter-based (P2MP) passive optical network type of Figure 1.3 and are based either on Gigabit Ethernet or **asynchronous transfer mode (ATM)**. We shall spend most of this chapter detailing the topologies, protocol layers, and message exchanges—in short, the architecture—of these two options. We shall rely almost entirely on the documents of the international standards bodies—the **Institute of Electrical and Electronics Engineers** (IEEE) for the Ethernet-based systems and the **International Telecommunications Union** (ITU), Section T, for those that derive from ATM.

In a PON there are only passive splitters between the central facility and the customer premises, no active devices, and in the rare case that amplification is required, it too is all-optical. Thus, a PON is a limited form of **all-optical network** [Green], one having no electronics except at the ends. PONs are all-glass trees, sending outgoing and incoming signals on different widely spaced wavelengths [coarse wavelength division multiplexing (CWDM), as distinguished from DWDM, D for dense]. DWDM systems have been widely deployed in telco and cable long-haul and metro communications, starting in 1991. Whereas these systems can use many tens or even hundreds of wavelengths at tight spectral spacing, today's PONs use only two or three widely spaced wavelengths.

Usually, the home run option is not called a PON, but since it has no electronics between head end and subscriber, in this volume we shall include home run topologies in our discussion. Some providers are finding this option attractive since long runs of multifiber cable are almost as economical in both material and installation costs as the same lengths of cables with one or a few fibers. The problem is the large electrooptic port count at the central office. However, there is the option of

Fiber to the Home: The New Empowerment, by Paul E. Green, Jr.
Copyright © 2006 John Wiley & Sons, Inc.

mitigating this by doing an optical split right at the CO, so that one optoelectronic port serves a number of subscribers in time-division multiplex (TDM) mode, whereupon the home run system becomes a PON in our usage.

The idea of DWDM PONs, in which each subscriber owns one of many wavelengths, appears in the literature from time to time (e.g., [Lee]) and might eventually prove attractive when component costs have hugely decreased from their present level. However, DWDM PONs have the huge drawback that they incur the high logistic costs of manufacturing, stocking, and provisioning frequency-stable laser diodes either tunable or fixed at a great many specified wavelengths. For the purposes of the present book, we shall not discuss them further.

■ 2.2 WHAT DOES A PON LOOK LIKE?

Figure 2.1 shows in more detail the typical PON being deployed today to provide one-stop shopping for the complete *triple-play* set of end-user services:

- Several RJ-11 twisted-pair *telephone* connections (**POTS—plain old telephone service**), served at the head end [also called the CO (central office) or **POP (point of presence)**] from one or more 155- or 622-MB/s G.303 traffic and control interfaces to the central office switch that is part of the public switched telephone network
- 10, 100, or even 1000BaseT Ethernet *data* services from RJ-45 connectors and Category 5 cable, served at the CO by one or more IP routers with IP over Ethernet interfaces
- Television distribution, either analog or digital, derived from satellites or microwave facilities

At the central office (head end), one optical line terminal (OLT) connects to the one-fiber or two-fiber tree structure, which in turn connects to the many optical network

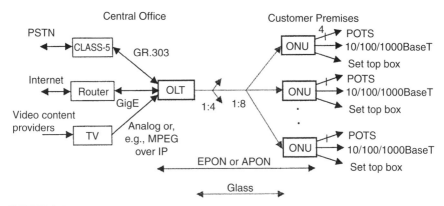

FIGURE 2.1 Typical residential passive optical network using a single bidirectional fiber for all of the triple-play services: voice, data, and video. The optical splits shown are merely illustrative.

units (ONUs). Although the option to have two fibers, one in each direction, is included in the architecture, almost all implementations have taken advantage of the cost savings available by using a single copy of the tree shown in Figure 2.1, with different wavelengths going and coming, so we shall not discuss them further.

With both voice and TV beginning to evolve toward the traditional data transmission mode, namely packet switching, the need for circuit-switched capability *anywhere* in the hierarchy is being questioned. It is becoming clear that the voice–data–TV triple-play service mixture of telcos and cable companies alike is tending toward packet switching based on IP. Many are suggesting that it is time to get rid of circuit switching even higher in the hierarchy, namely upstream from the CO, the backhaul part of the system. Referring to Figure 2.1, there is a movement, sometimes referred to as "POTS to pipes," to get rid of the expensive Class 5 Signaling System 7 circuit switch, driven across the GR.303 interface, and to carry all three forms of traffic through routers [Isenberg, Ross]. To reconfigure an SS-7 switch is estimated to cost tens of dollars, but to forward an IP packet costs only a few cents [Ireland].

Not shown in Figure 2.1 is the fact that, for PON service to some businesses, part of the capacity otherwise going for television is replaced by T1 and T3 appearances to serve customer private branch exchanges (PBXs). Also, larger businesses tend to format their communication resources into virtual private networks (VPNs), today's (often IP-based) successor to earlier enterprise networks of leased private voice-grade and T-carrier lines.

Originally, any PON based on ATM was called an **APON**, but the term became ambiguous when two distinct forms emerged. These are the **broadband passive optical network (BPON)**, which conveys only cells and the **gigabit passive optical network (GPON)**, which conveys either cells or frames, or both **(mixed mode)**, both standardized by ITU-T. All of the ITU-generated PON types (BPON and GPONs in their various options) are sometimes referred to collectively by the name of the consortium that originated ATM-based PON standards and then turned them over to the ITU: **FSAN (full service access network)**. The Ethernet-based passive optical network has always been called an **EPON**, although sometimes one hears the term **EFM (Ethernet in the first mile)**, a term that more accurately describes any use of Ethernet protocols for access, not just EPONs. The gigabit per second flavor of EPON is also known as **G-EPON**.

For BPONs, GPONs, and EPONs alike, one thing that the ITU and IEEE have agreed on is that downstream binary data, voice or video, will travel at 1490-nm wavelength, upstream digital data, voice or video, at 1310 nm, and analog video will travel downstream at 1550 nm. Analog video presents so many special technical problems when transmitted optically that we shall have a long discussion of it when describing the BPON standard, the only one of the three to deal with it in any great detail to date.

The PON standards specify upper limits to the number of subscribers that can be connected to a PON, but in practice the number is limited by the available signal-to-noise ratio (SNR), and something called the laser turn-on delay that we shall discuss in the next chapter. Architecturally, the number of bits available in addressing the subscribers is usually much more generous than what the standard says about number of stations.

■ 2.3 ATM CELLS OR ETHERNET PACKETS?

Asynchronous transfer mode (ATM) [DePrycker], a form of packet switching, was at one time the great white hope of the world's telephone companies for providing a total repertoire of digital services. This hope still echoes today in the race to provide "triple play," except that today it is a race with the cable providers instead of with the computer community.

It had become clear that circuit switching had to give way to packet switching, or at least switching of virtual circuits (VCs), where each VC was a flow of identically addressed packets. The problem in adopting packet switching as practiced then by the computer community, was thought to be the delay-sensitive nature of voice traffic exacerbated by the need to buffer this fairly low-bit-rate traffic in a concatenated series of COs, a hard lesson that had been learned from abortive attempts to substitute point–point synchronous satellite links (with a 0.25-s round-trip delay per hop) for terrestrial links. So the phone companies and their standards bodies opted for extremely short 53-byte packets, "cells," with payloads of only 48 bytes. This choice was made in spite of the high header overhead, the clearly impending widespread availability of much higher transmission bit rates along with much lower bit error rates leading to fewer retransmissions, the fact that non-voice traffic was going to longer and longer packets as storage got cheaper and applications became more bit-hungry, and the fact that data bit rate volumes would soon exceed POTS bit rate volumes. The dominance of voice in the telco revenue stream was allowed to dictate exclusive use of cells in the architecture of ATM.

The computer fraternity watched this choice of a mere 53 bytes in befuddlement, wondering why, in view of ongoing telecomm evolutions, particularly higher speed on long-haul trunks, one wanted such short transmission quanta. There were many ideological debates over the idea of basing all of future telecommunications on such a choice. Today, the "great cell debate" has receded in importance for several reasons:

- Nobody except the phone companies uses cells, and they use them only for voice and internal network control. The predicted cell-based cornucopia of voice, image, sensor, and other applications simply never evolved, actually not just because they were to be cell-based but simply because those innovations that succeeded tended to come from Silicon Valley and the computer fraternity, not from the telecommunications carrier fraternity.

- Bandwidth became so plentiful that having packets that were so short that the header overhead was a whopping 8% no longer bothered many people. Similarly, microcomputers became sufficiently powerful that servicing an interrupt every 53 bytes proved easier to handle than had been the case earlier.

- GPON will eventually supersede BPON because the cells can be made somewhat submerged and invisible. Starting in 2001, the ITU defined a way of hiding the underlying existence of 53-byte cells by enveloping multiple cells into more reasonable large packet sizes called **GPON encapsulation method (GEM)** frames, a feature of GPON. The inherited 125-μs framing remains, but the cells can be optioned out. EPON has essentially no packet length restrictions.

This means that as soon as the telcos that are offering FTTH services move off BPON and onto GPON, the remaining incentive for EPON is that GPON's complexities and 125-μs packet duration limit add somewhat more cost, as we shall discuss in making a detailed comparison of the three in Section 2.8. While reliable delivery of traffic in the form of short, fixed-length ATM cells might be a natural and widespread practice within the internals of the world's telecommunication carriers, the world of end users has found it more natural to speak in long packets of user-defined lengths, not telco-defined lengths. BPON speaks in cells, GPON in both cells and in packets no longer than 125 μs (frames), and EPON in user-defined packets.

The standard that has covered more of the issues than the other two is that for BPON, as we shall see. This is because of its relative maturity. What the EPON standard has going for it, in competition with GPON, is the very mature learning curve on component cost reduction and user acceptance. ATM has been a telco favorite, with corresponding attention to high-quality (and therefore expensive) engineering. Ethernet, on the other hand, can be bought at the corner stationery store, and this has been true for years.

■ 2.4 HOW THE ARCHITECTURES WILL BE PRESENTED IN THIS BOOK

For purposes of this book, specifying a communication architecture standard, and doing so completely, consists of giving three different views of it. First one splits all the functions within each of the two communicating partners (e.g., an OLT and an ONU) into *layers* in a protocol stack such that, at any given instant, a layer instance in one node talks over the network only to a single distant partner instance of the same layer in the other node. At the same time it is talking and listening both upward and downward to the adjacent layers in its own node. Then, one must describe the *messages* that the two peer members at a given level in the stack exchange with each other. Finally, one must discuss the *sequence of events* that each message exchange produces.

For completeness one should define the layers, formats, and message exchanges not only for steady-state message exchange but also of the overall transitory supervisory functions. This means that, for completeness, there should be two protocol stacks, one for steady-state message exchange, the **data plane**, and the other for the intermittent control functions, the **control plane**. The standard two-plane Open System Interconnection (OSI) view is shown in Figure 2.2.

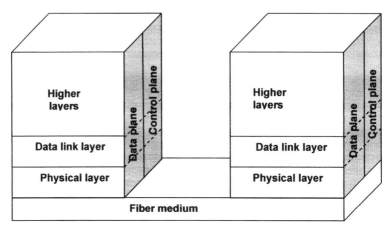

FIGURE 2.2 Division of the architecture of each of two communicating nodes into two planes, each constituting two protocol stacks: the data plane and the control plane.

More specifically, the three views of a protocol stack are:

- Portrayal of the structure of **layers, sublayers**, and so forth with a word description of what each layer does. The data plane of layers, delivers the steady-state traffic to and from the end user, in this case, the port at the OLT and at each premises ONU. The control plane, or **operations, administered and management (OAM)** or **operations support system (OSS)** plane of layers, embodies the operations, control, and management functions. These intermittently invoked OAM functions involve such things as initial setup, error recovery, situation reporting, and, in the optical case, such special things as adjusting laser power while traffic is flowing. In this book, we shall give minimal treatment to the OAM part of the architecture, concentrating on the flow of traffic in a healthy already running network.

- Enumerating what each **header, subheader**, and so forth (usually preceding the user traffic fields) contains in the way of control information, it being understood that header information belonging to one layer will not even be seen by layers above it at the sender or receiver. If we have enumerated, bit by bit and byte by byte, what the formats of the messages between these cooperating peers look like, we have defined their **syntax**. One can learn a great deal about what the protocol does just by studying what a node says and what it hears.

- Giving complete timelines or **state diagrams** for the permitted message exchanges between peers, the exchanges of commands and responses between adjacent levels vertically in each of the two protocol stacks, plus the defined actions taken upon receipt of messages horizontally and vertically. Specification of these things constitutes the **semantics** of the architecture.

For present purposes, we shall be content to do some of all three for the data plane. We shall hardly touch on functions in the control plane. For details on this, the reader should consult the standards documents we shall discuss or the Telcordia GR-909 document [GR-909].

As for the semantics, while most of the content of the standards documents deal with these detailed sequence diagrams and state tables, we shall leave much of this semantic detail for the specialized reader, hitting instead only a few high spots.

For BPONs, then GPONs, then EPONS, we first give a sequence of exploded views of the layers and sublayers, then a sequence of exploded views of the headers, subheaders, and so forth, and finally selected parts of the mass of detail that the standards documents give on the protocol semantics.

As to the semantics, we particularly need to say something about:

- The **ranging** protocol, a frequent message exchange by which the OLT measures the propagation time to the various downstream ONUs, so that their transmissions will be time interleaved in a nonoverlapping way.

- **Dynamic bandwidth allocation**, which allows the reallocation and reuse of otherwise unused upstream capacity.

- Measures to provide **security**, obviously an important matter in a topology where any ONU is exposed to the downstream traffic from the OLT to all ONUs. The ability of one ONU to eavesdrop on another's *upstream* traffic is much less of a problem, unless the splitters or the OLT receiver has a significant back-reflection (low return loss).

- **Protection switching**, the substitution of a new optical path for a failed one, which is allowed for with the BPON and GPON standards, but not EPON.

■ 2.5 ITU'S BPON (BROADBAND PASSIVE OPTICAL NETWORK) STANDARD G.983

2.5.1 BPON Portrayed as Layers

Figure 2.3 shows the basic BPON layer structure [G.983.4]. The subscriber's ONU has one instance of all the communication-level layers, whereas at the head-end OLT there are many instances of some of the layers equal to the number of active ONUs. The bottom layer, the one that interfaces to the glass part of the network—that is, the **optical distribution network (ODN)**—is the **physical medium layer.** This layer takes care of electrical-to-optical conversion, directing signals to or from the medium at one of the three optical wavelengths (1310, 1490, or 1550 nm), and connection to the fiber of the ODN. The syntax of this layer consists simply of a complete statement of all the standardized optoelectronic parameter choices.

Between the physical medium layer and the interface over which ATM cells are delivered to the client layers is the **transmission convergence layer** (corresponding roughly to the OSI level 2 of Fig. 2.2). The TC layer is subdivided into the PON **transmission sublayer**, and above this the standard **adaptation sublayer** inherited from traditional copper-based ATM [ITU I.732]. This layer's function is to convert between user-level 125-μs frames (**protocol data units—PDUs**) and ATM cells.

Path layer			ATM cells to user frames
Transmission convergence layer	Adaptation		BISDN network-to user frames

	PON transmission		Ranging
			MAC
			   Cell slot allocation base on
			      DBA
			   Detection of offered load
			      for DBA
			   BW allocation based on
			      traffic contract and BW
			      request
			Privacy and security
			Frame alignment
			Burst synchronization
			Bit/byte synchronization

Physical medium layer			E/O adaptation Wavelength division mux. Fiber connection

FIGURE 2.3 Protocol stack for data flow in a BPON. The stack for control (initiating, terminating, and recovery of traffic flows) is not shown.

BPON Physical Medium Dependent Layer

This layer, unlike the higher layers, is composed of hardware, not software. This hardware is defined by the standards [G.983.1, G.983.3] as obeying the following parameters:

- Bit rates: 155.52 or 622.08 Mb/s down and 155.52 up
- Wavelengths: 1260 to 1360 up, 1480 to 1580 nm down
- Traffic type: Digital bidirectional and downstream analog
- Fiber splits: Up to 32, limited by ODN attenuation
- Worst-case attenuation permitted between the OLT and the ONU (the ODN). The worst-case qualifier includes not only bulk fiber attenuation but also losses in splices, connectors, optical attenuators (if any), or any other passive devices, plus a safety margin to cover future additional splices or added cable length, change of attenuation with environmental factors, and possible loss degradation at the connectors:

Class B   10 to 25 dB

Class C   15 to 30 dB

- Maximum fiber distance: 20 km, as set by attenuation and the uplink protocol.
- Launch power range for one-fiber case and 0.622 Gb/s downstream in dBm:

Class B and C -2 to $+4$ dBm

- Same for 0.155 Gb/s upstream:

  Class B   −4 to +2 dBm
  Class C   −2 to +4 dBm

- Launch power during slot not assigned to this ONU: Less than 0.1 of "1" bit level.
- Worst-case downstream receiver sensitivity/minimum receiver overload at 622 Mb/s in dBm:

  Class B   −28/−6
  Class C   −33/−11

- Worst-case upstream receiver sensitivity/minimum receiver overload at 155 Mb/s in dBm

  Class B   −30/−8
  Class C   −33/−11

- Maximum bit error rate: 10^{-10}
- Laser extinction ratio (either direction): >10 dB
- Laser linewidth at −20 dB from peak: Downstream DFB laser: 1 nm max. Upstream **MLM (multiple longitudinal mode** otherwise know as Fabry-Perot) laser at 155 Mb/s: 5.8 nm. For upstream **SLM (single longitudinal mode)** lasers, usually distributed feedback (DFB) lasers at all bit rates: linewidth of 1 nm max.

BPON PON Transmission Sublayer of the Transmission Convergence Layer

This layer deals entirely with cells. Inbound, they are derived from the electrical signal delivered by the physical medium layer, synchronized at bit and byte level, the cell and frame boundaries determined, the header stripped off and processed, and each individually addressed cell stream delivered to the appropriate instance of the next higher layer. Outbound, the process is reversed. It is in this layer that the ranging protocol is periodically executed to ensure that cells from the various ONUs do not overlap. Two other important functions that we shall discuss also live here, dynamic bandwidth assignment (DBA) and encryption.

BPON Adaptation Sublayer of the Transmission Convergence Layer

This is the layer where the conversions take place between ATM cells and PDUs, which include such things as SONET/SDH, xDSL, and other *telephone company* protocol data units that are based on 125-μs framing [ITU I.732]. It does not provide a native interface for such packet-based traffic as Ethernet or IP. These needs must be accommodated by added protocol conversion software that is outside the scope of the standard. As we shall see, the ITU finally addressed this need with GPON in a way that did not require such nonstandardized protocol conversions, but it did require the user packets to be broken up if they were longer than the standard telco 125-μs frame.

2.5.2 BPON Portrayed as Formats

The frame formats for the BPON are quite simple. As Figure 2.4*a* and 2.4*b* show, the 125-µs frame formats are different for 155- and 622-Mb/s downstream transmission. The upstream 155-Mb/s frame format is independent of which downstream bit rate is in use. From comparing Figure 2.4*a* and 2.4*b*, it is seen that the only difference in format between 155- and 622-Mb/s downstream service is the number of downstream PLOAM/data cell pairs occupying one 125-µs frame.

Each *downstream* frame consists simply of a sequence of 53-byte control cells called **physical level operations and maintenance (PLOAM)** fields, followed by 27 cells of user data, followed by another PLOAM, and so forth. Each 53-byte cell downstream is also called a **time slot** in the standards documentation, but the term is confusing since upstream a time slot is 56 bytes, as we shall see. To add to the confusion, the name PLOAM is a little confusing too because the PLOAM

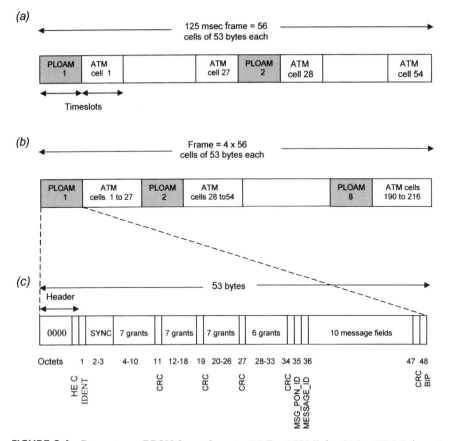

FIGURE 2.4 Downstream BPON frame formats: (*a*) For 155 Mb/s, (*b*) for 622 Mb/s, and (*c*) details of the PLOAM field.

originates and terminates in the PON transmission sublayer of Figure 2.3 not in the physical medium layer, as the term "physical level" might lead one to believe.

The PLOAM cell is the heart of the PON's operation, and this is true in both directions. In the downstream direction, when upstream dynamic bandwidth assignment (DBA) is implemented, it is the **grants** that are most important—permissions for each ONU to send one or more cells in succession to form a T-Container (T-Cont). In the upstream direction the most important PLOAM content is the diagnostic information that the ONU gives the OLT, reporting its own state of health.

Let us complete the discussion of downstream BPON syntax by picking apart the downstream PLOAM and then the upstream version. Figure 2.4*c* shows a blowup of the downstream PLOAM, identical for both 155 and 622 Mb/s.

- A five byte header of which the first four bytes are all set to fixed bit patterns, and the remaining byte is the **header error correction (HEC)** byte, which serves to provide the very heavy protection to the header that it needs to make very sure that the start of the PLOAM is identified with certainty.
- The IDENT field says whether this is the first PLOAM in the downstream frame or not.
- The two bytes of frame sync can be used optionally to deliver from the OLT to the ONU a 1-kHz reference signal.
- Most of the rest of the PLOAM consists of 27 grants, each set of six or seven protected by its own **cyclic redundancy check (CRC)**. A grant tells a given ONU when it can send during its next frame either a T-Cont (succession of ATM data cells) or a PLOAM cell, or nothing. The ONU can also be instructed here to rerun the ranging process.
- MSG_PON_ID tells a particular node in the PON that the following message is for it, and MSG_ID tells the message type.
- Ten bytes of CRC-protected messages, which the OLT uses to tell the ONU its proper ranging offset, and which the OLT can also use for various implementation-dependent purposes, for example, whether to readjust power level or extinction ratio.
- The **bit-interleaved parity (BIP)** byte that contains the parity of all the bytes transmitted since the last BIP, allowing the ONU to monitor the bit error rate.

Two options are available for the use of *upstream* capacity: static (fixed) allocation or dynamic bandwidth allocation (DBA). It is possible to operate both options concurrently. A fixed assignment upstream frame is shown in Figure 2.5*a*, each frame consisting of fifty-three 56-byte time slots, each consisting of one 53-byte ATM cell, preceded by an implementation-specific 3-byte overhead consisting of:

- At least 4 bits of **guard time** to avoid collisions in case of some inaccuracy in the ranging protocol that sets up the proper interleaving, to be described.

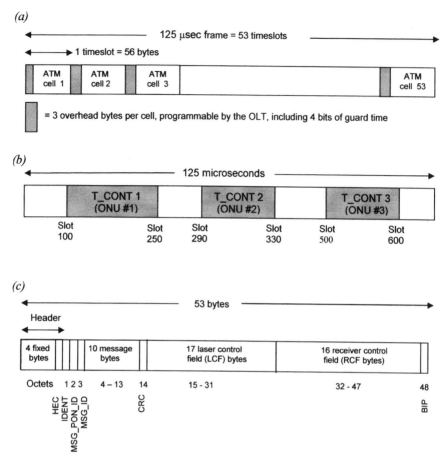

FIGURE 2.5 Upstream BPON frame format: (*a*) The fixed case—transmission as individual ATM cells, (*b*) the DBA case—transmission as series of identically addressed long cell strings called T-Conts, and (*c*) use of one cell duration for a PLOAM cell.

- A **preamble** pattern that allows the OLT to acquire bit timing and read the amplitude.
- A **delimiter** for byte sychronization and/or to indicate the start of the ATM cell.

Although Figure 2.5*a* shows all the 53-byte fields being completely occupied by ATM user cells, under command of the OLT, any ONU can use its assigned 53-byte cell slot time to send either a 53-byte PLOAM cell or its portion of a divided slot, not shown in Figure 2.5*a*.

In 2001 the ITU introduced dynamic bandwidth assignment to the BPON architecture [G.983.4]. This protocol is often also referred to as the BPON **media access control (MAC)** protocol. It still did not allow the PON to transport packets, only

ATM cells, but at least the number of cells per frame per user ONU could be changed dynamically by the OLT and sent upsteam by the ONU in bursts. In DBA, the OLT allocates units of upstream capacity called **transmission containers (T-Conts)** to the various ONUs so that each ONU can then send a string of cells within the duration permitted by the OLT, namely the duration of the T-Cont. The DBA mechanism provided a capability to share the upstream bandwidth in a dynamic and flexible way.

Figure 2.5*b* shows another view of the upstream 125-μs frame of Figure 2.5*a*, relabeled to show how a succession of T-Conts is interleaved (each composed of time slots, each in turn containing a cell), one from each of three ONUs. Each T-Cont flow from ONU to OLT can accommodate one or more **virtual paths**, and each VP can accommodate one or more **virtual circuits.** VCs and VPs [Tanenbaum] are entities inherited from earlier practice, going all the way back to X.25 in the 1970s.

Note that the term *time slot* means a different thing downstream and upstream. Downstream, a time slot is one 53-byte cell. Upstream it is a cell plus a 3-byte overhead field.

The upstream PLOAM syntax is shown in Figure 2.5*c*:

- As with the downstream case, it starts with the same 5-byte header and 1-byte of IDENT, whose contents are reserved for future use.
- Similarly, the MSG_PON_ID and MSG_ID bytes serve to identify the ONU, in this case the one sending.
- Ten bytes of CRC-protected messages carrying alarms, threshold crossings, and other OAM information from the ONU.
- Seventeen bytes of **laser control field (LCF)**, used for the ONU to report the power level and extinction ratio of its own laser so that they may be maintained at the proper values.
- Sixteen bytes of **receiver control field (RCF)**, used by the OLT's receiver to correct its own detection threshold level for distinguishing a zero from a one.
- The bit-interleaved parity (BIP), whose function is the same as with the downstream case.

There is a third alternative to an upstream ATM cell or PLOAM cell, namely something called a *divided slot*. The upstream divided slot is what makes dynamic bandwidth allocation work. One divided slot consists of a number of *minislots.* Each minislot in the overall 53-byte divided slot is transmitted by one particular ONU, so that many can contribute their minislots within one slot time. Each minislot begins with the same 3-byte overhead that precedes a cell or PLOAM, and each minislot ends with an 8-bit cyclic redundancy check to protect it individually. In the interval between the overhead and the concluding CRC, the minislot tells the OLT its T-Cont status. The MAC protocol at the OLT uses the information in the minislots from individual ONUs to then execute the dynamic bandwidth allocation protocol reassigning bandwidth quanta to each.

2.5.3 BPON Portrayed as a Sequence of Events

It has been fairly easy to present the BPON layers and syntax (views of header fields and subfields), but the semantics are another matter. To get a picture of the BPON architecture specification detailed enough for one to troubleshoot a system or sit down and start writing code would require a very large amount of material that is covered in helpful detail only in the ITU's G.983 documentation [G.983.x]. However, for the nonspecialist reader, a few highlights of what is going on are much more helpful and manageable in a book of the present scope. Rather than try to give detailed timelines or state diagrams of all the initiation, takedown, and recovery actions of the many substeps in operating a PON, we shall confine the discussion here to the three topics mentioned earlier: ranging, security, and protection switching.

2.5.4 Ranging

It is clear that when the many ONUs are all transmitting toward the OLT matters must be arranged so that their frames cannot overlap each other on arrival. They must lie in the pattern shown in Figure 2.5a or 2.5b. Since the various ONUs may lie at various distances from the OLT, each ONU must learn just when to launch its next frame so that it will arrive properly time interleaved with the frames from other ONUs. Outbound there is no problem, since the OLT is in charge and can itself properly time interleave the variously addressed frames. In order for inbound transmission to take place, the measurement of range must be made when the ONU is first powered on and must be repeated periodically thereafter because of possible aging and temperature effects on the propagation time. At startup, the OLT can assign to each ONU an initial range between 0 and 20 km, after which the ranging protocol adjusts the time delay in the ONU to its correct value. Thereafter, to start the ranging process, the OLT issues a *ranging grant* as part of the downstream PLOAM cell addressed to the station to be ranged and opens up a time window long enough to receive the upstream ranging PLOAM.

Once the OLT knows the range to each ONU, it notifies each ONU of its assigned **equalization delay**, which then gives the ONU a window within which it is allowed to start sending. The ranging process that is used maintains the range knowledge at the OLT to amazing accuracy, only 8 bits at 1.2 Gb/s (a GPON bit rate), or 4 bits at 622 Mb/s or 1 bit at 155 Mb/s with BPON.

Ranging is initiated by the OLT at repeated intervals after an initial success, and also polls from time to time to see if any new ONU's have been added. Both these intervals are programmable, not standardized.

2.5.5 Security

Two mechanisms are provided with BPON for ensuring that traffic downstream is read only by the intended ONU:

- *Churning*, in which all downstream cells to an ONU are permuted in a pseudorandom way, and the permutation key changed once per second.

When commanded by the OLT, the ONU sends upstream a data-dependent 3-byte key that the OLT then uses to permute (churn) subsequent downstream cells.

- While *encryption* of at least the user data fields has been left out of the BPON standard, it has been included for GPON, so we shall mention it there. For BPONs, encryption remains a user-defined option.

One last remark about security. Since the upstream data travels at its own wavelength, 1310 nm, and is time interleaved and since there will always be back-reflections from each splitter, there will be a tiny replica of every ONU's transmission upstream that could, in principle, be detected and analyzed at another ONU. No numbers are given on this possibility in the standards documents, only the specification that the return loss (echo at the ONU from splices, splitters, and connectors) must be at least 32 dB down from the upstream power level. It is unstated whether an out-of-spec component giving less than 32 dB return loss or an out-of-spec 1310 laser delivering more power than specified can compromise a user's security.

2.5.6 Protection Switching

Automatic protection switching is a physical level redirection of all the data and control from one path to another. Building on the success of automatic protection switching in protecting SONET/SDH facilities by providing instantaneously available disjoint backup paths (e.g., in the two counter-rotating directions of a ring), the ITU has provided at least a place holder few paragraphs in the BPON [G.983.1] and GPON [G.984.1] standards documents. Even though the switchover is typically done in milliseconds, data and state information must be retained and recovery instituted, and so protection must properly be considered part of the function of the OAM plane of Figure 2.2 rather than the normal data flow processes. This is in contrast with the ranging, dynamic bandwidth assignment, and security functions we have just been discussing, which are, to a great extent, ongoing steady-state functions.

There are several possible geometries, depending on which portion of the system is considered most vulnerable, the OLT ports, ONU ports, the entire glass ODN, or just the long trunk leading to the first optical splitter. In some of these cases the switchover is electronic, being before a pair of OLT ports and/or after a pair of ONU ports. In other cases, the switching must be photonic.

Figure 2.6 shows three of these possibilities [G.983.1]. The disadvantages of Figure 2.6*a* include the fact that the switching must be photonic and that, because of a signal loss during switching and the difference in geographical distance along the two paths, loss of one or more frames is possible. With Figure 2.6*b* there is a hot standby OLT electrooptic front end and two disjoint fiber trunks out to the first optical splitter. Even if the backup OLT front end has a time offset to compensate for the difference in distance, signal and possibly frame loss is inevitable. If truly **hitless switching** (no loss of signal or framing) is absolutely required, it can be achieved with the arrangement in Figure 2.6*c* in which the entire cascade of OLT front-end, ODN, and ONU front-end portions is duplicated. The switchover at both ends is done electronically between, in effect, two live operating PONs already synchronized and carrying the same live traffic. Geographical path length

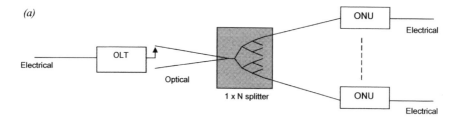

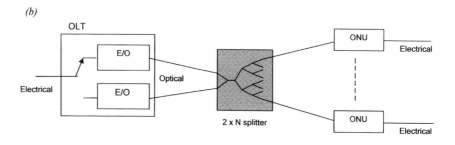

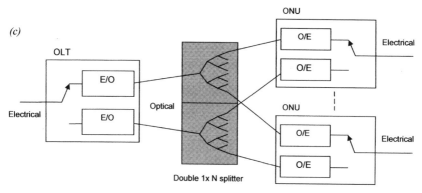

FIGURE 2.6 Several PON protection switching schemes: (*a*) photonic switching of trunk out to first splitter, (*b*) duplexing of trunk and OLT front end, and (*c*) duplexing of entire cascade of OLT back end, ODN and ONU front end.

differences will already have been built into the operating state of both main and backup PON facilities.

As PONs begin to be more widely installed, not just for residences, but for small and then for larger businesses, one may expect the ITU and IEEE to turn their attention again to this question and work out further details.

2.5.7 Analog Video Delivery over a BPON

While the BPON architecture allows for analog delivery of video, both GPON and EPON documents assume, at least implicitly, either that the video part of the payload will be digital or that the BPON document has standardized all that needs to be standardized. In the BPON architecture, the 1550-nm wavelength region has been reserved for an **overlay** of analog video [Perkins] on top of the conventional 1310 for data upstream and 1490 for downstream.

Video dominates, and is likely to continue to dominate, the outbound payload size and therefore the architecture and componentry of optical broadband access. Moreover, video will some day become an important traffic component going inward toward the OLT. The impending massive arrival of HDTV only serves to exacerbate the already identifiable dominance of TV over voice and data traffic volumes.

Today it is *analog* TV that most of the world uses, for one thing because almost everywhere it is free and those channels that are not digitized (e.g., using QAM) require no set-top box. For present purposes we shall include in our definition of analog such mixed analog/digital systems as those that send many frequency-multiplexed channels that are individually QAM digital signals—the ultimate character of the transmitted mix is an analog sum.

Sales of analog TVs continue to be strong, while it is estimated that the average consumer keeps a TV set for an average of 15 years or more, and therefore will for at least that long expect his service provider to cater to that need. As we shall see in the next chapter, analog TV gobbles up the most promising wavelength band in the fiber and requires orders of magnitude more precious received optical power. This in turn leads to much tighter crosstalk tolerances on wavelength division components, while these same high-power levels produce a variety of strange propagation effects that must be carefully dealt with. So the evolution of TV broadcast from analog to digital has very important payoffs in terms of PON system first cost, lifetime maintenance costs, and room for growth.

Eventually, the replacement of analog TV by digital IPTV will occur. However, given the huge installed base of analog TV, 15 years does not seem an unreasonable estimate for the time it will take to migrate TV over to digital in the developed countries. In the developing countries it could take longer.

Present State of FTTH Analog Video Standardization

Certainly one deterrent to standardization of the 1550 overlay has to do with the "tower of Babel" nature of entrenched television practices internationally, with different regions (or countries within a region) having different standards on pixels per line, lines per frame, frame interleaving, frame rate, and color encoding [Poynton]. Only the upgraded BPON document [G.983.3] of 2001 has anything at all to say about analog TV, and not very much at that. Unfortunately for the guidance of PON implementers, "optical parameters, line code, modulation scheme and so on for the enhancement band are out of scope of this recommendation." This leaves all of today's PON implementers to create their own ad hoc parameter choices or rely on partially frozen solutions being debated within the standards bodies. What the G.983.3 document does include (in its Appendix III) are some sample calculations of the permitted power level limits and rejection ratios in WDM components necessary to isolate video services from data services and vice versa in the presence of significant power levels and PON optical impairments.

It appears that with BPON the ITU has gone as far as it will go in catering to analog TV needs and has now, with GPON, turned its attention to "leading edge" questions such as gigabit digital rates, where, by implication, video is included as part of the triple-play mix.

Important FTTH Analog Video Impairments

In Section 3.2 we shall discuss the effect of optical nonlinearities and other afflictions in the actual fiber propagation medium. These can lead to loss of SNR at the optical receiver, instabilities in the transmitted power level, and contamination of signals in one of the wavelength bands (1310, 1490, and 1550) by power in one of the others. With analog video downstream at 1550 nm, we face exactly this problem, unlike the digital video case where the video stream is part of the overall OLT-to-ONU bit stream at 1490 nm, which is at lower power.

Not only are there tiny nonlinearities in the fiber medium that can accumulate to surprising levels, but there can easily be nonlinearity in the relationship between input to the laser at the OLT and output from the photodiode at the ONU. For analog video, great care and considerable expense must be devoted to making this input–output relationship as linear as possible for reduction of crosstalk, particularly for multi-channel FDM analog video. Also, great pains must be taken to reduce received noise.

The three parameters of the signal delivered to the TV receiver that define the relevant impairments for video are:

- *Carrier-to-noise ratio* (*CNR*), another name for signal-to-noise ratio.

And two intermodulation terms:

- *Composite second-order* (*CSO*) terms
- *Composite triple beat* (*CTB*)

Composite second order and CTB arise in the following way [Green, Section 9.3]. Imagine that a number M of simultaneously present sinusoids at equally spaced frequencies f_i ($1 < i < M$) (representing M video channels) are passed through a nonlinear device or nonlinear cascade of devices. There will be second-order intermod terms at frequencies ($f_i \pm f_j$) and third-order terms at ($f_i \pm f_j \pm f_k$), where i, j, and k range from 1 to M. The second-order terms can be ignored since they lie either near direct current (DC) or at double frequencies, but certain third-order terms constitute CSO and CTB. CSO terms are those of the form ($f_i + f_j - f_k$) with $i = j$ (they are termed "second order" because there are only two distinct frequencies). CTB terms are those of the same form ($f_i + f_j - f_k$), but with i, j, and k all different. CTB can be much stronger than CSO because there are $M/2$ times as many instances.

Poor CNR appears as snow on the picture, while CSO and CTB, indicating the level of coherent beats between TV channels, appear as lines and stripes. While the G.983.3 BPON standard sets limits on CNR, but not on CSO or CTB, the Federal

Communications Commission has done so for all three [FCC], setting minima of 43 dB for CNR (as has ANSI [ANSI/SCTE]), and −51 dB relative to the carrier for both CSO and CTB. There are very aggressive numbers, as we shall see in the next chapter.

■ 2.6 ITU'S GPON (GIGABIT PASSIVE OPTICAL NETWORK) STANDARD G.984

2.6.1 GPON Portrayed as Layers

As shown in Figure 2.7*a*, the layer structure of a GPON OLT or ONU consists, at the grossest level, of two layers: the **physical-medium-dependent (PMD) layer**

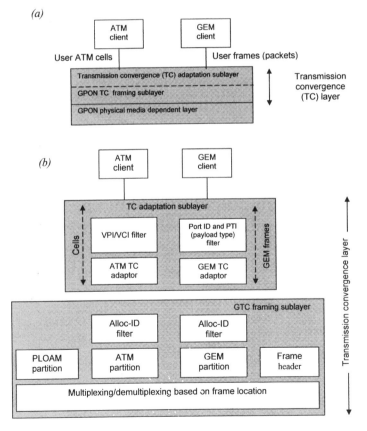

FIGURE 2.7 Protocol stack for data flow in a GPON node (the stack for control—initiating, terminating, and recovery of traffic flows—is not shown). (*a*) All the layers and (*b*) exploded view of the adaptation and framing sublayers.

[G.984.1] and the **transmission convergence (TC)** layer [G.984.3]. They correspond roughly to the OSI physical layer and data link layer of Figure 2.2, respectively. The TC layer is in turn divided into the **framing sublayer** and the **adaptation sublayer**. We discuss first the physical-medium-dependent layer of Figure 2.7a, then the framing sublayer of Figure 2.7b, and then the adaptation sublayer of Figure 2.7b.

GPON Physical-Medium-Dependent Layer

This layer, unlike the higher layers, is composed of hardware, not software. This hardware is defined by the standard [G.984.2] as obeying the following parameters:

- Bit rates: 1.24416 or 2.48832 Gb/s down and 0.15552 or 0.62208 or 1.24416 or 2.48832 Gb/s up
- Wavelengths: 1260 to 1360 up, 1480 to 1500 nm down
- Traffic type: Digital only
- Fiber splits: Up to 64, limited by ODN attenuation
- Attenuation permitted between the OLT and the ONU (the ODN). The worst-case qualifier includes not only bulk fiber attenuation but also losses in splices, connectors, optical attenuators (if any) or any other passive devices, plus a safety margin to cover future additional splices or added cable length, change of attenuation with environmental factors, and possible loss degradation at the connectors:

  Class A: 5 to 20 dB
  Class B: 10 to 25 dB
  Class C: 15 to 30 dB

- Maximum attenuation difference between ONUs: 15 dB
- Maximum fiber distance: 20 km with DFB laser upstream, 10 with Fabry-Perot
- Launch power range for 1-fiber case and 1.2 Gb/s downstream in dBm:

  Class A: −4 to +1
  Class B: +1 to +6
  Class C: +5 to +9

- Same for 2.4 Gb/s downstream:

  Class A: 0 to +4
  Class B: +5 to +9
  Class C: +3 to +7

- Same for 1.2 Gb/s upstream:

  Class A: −3 to +2
  Class B: −2 to +3
  Class C: +2 to +7

- Minimum downstream receiver sensitivity/minimum receiver overload in dBm:

At 1244 Mb/s		At 2.488 Mb/s	
Class A:	$-25/-4$	Class A:	$-21/-1$
Class B:	$-25/-4$	Class B:	$-21/-1$
Class C:	$-26/-4$	Class C:	$-28/-8$ (requires APD receiver)

- Minimum upstream receiver sensitivity/minimum receiver overload at 1.2 Gb/s in dBm:

Class A:	$-24/-3$
Class B:	$-28/-7$
Class C:	$-29/-8$

- Launch power during slot not assigned to this ONU: Less than 0.1 of "1" bit level.
- Maximum bit error rate: 10^{-10}
- Laser extinction ratio (in either direction): >10 dB
- Laser linewidth at -20 dB from peak: Downstream DFB laser: 1 nm; upstream MLM (multiple longitudinal mode or Fabry-Perot) laser at 155 or 622 Mb/s: 5.8 and 2.7 nm, respectively, and not recommended for higher rates. For upstream SLM (single longitudinal mode or distributed feedback, DFB) lasers at all bit rates: 1 nm.

GPON Transmission Convergence Layer

In GPON (and BPON) parlance, a frame is a 125-μs quantum, inherited from earlier telco practice with T-carrier and SONET. There are two flows of GPON user frames into and out of a node from the PON optical distribution network (the glass infrastructure), whether it is an OLT or an ONU, namely flow of frames containing ATM cells, and **GPON encapsulation method (GEM)** frame flows. As we saw, BPON carries only one type of user traffic flow, namely cells. The cells can be catenated into burst (T-Conts), but they are still cells.

The GEM flow consists of long 125-μs GPON frames, whose length may or may not match that of the client user. If they do not, encapsulation is required. The word refers to the process of placing a user packet that might be shorter than the GPON frame within the latter, or if the user frame is longer than the GPON frame, of breaking it into fragments that are transmitted, fragment by fragment in successive GPON frames. The options are shown in Figure 2.8. In this GPON discussion, since we are not emphasizing ATM cells that propagate all the way to or from an end user, but only as a control function or to or from a real piece of telco ATM equipment, we shall discuss only the GEM function, not the cell function.

In the outgoing direction in a node, frames are put together in the GEM partition of Figure 2.7*b*. A header is prepared and appended that includes the **embedded OAM**, those few control functions that flow with every packet, and are therefore not really so intermittent that they are properly regarded as flowing in the control plane in Figure 2.2. These especially include dynamic bandwidth assignment

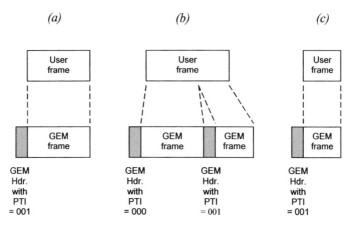

FIGURE 2.8 Encapsulation of downstream user frames into GEM frames: (*a*) user frame matches GEM frame, (*b*) user frame longer than GEM frame, and (*c*) user frame shorter than GEM frame.

(DBA), the requesting and granting quanta of bandwidth on a dynamic basis, as was explained for BPON.

In order that the system handle at the same time a mix of pure ATM 53-byte cells and also long packets, it partitions the downstream payload part of each frame into an ATM section and a GEM section, as we shall see in the next section. In the upstream direction, traffic is carried in T-Conts (transmission containers). Any T-Cont can carry ATM traffic only (as was the case with BPON) or GEM frame traffic only—a mix of both is supported by having some ATM T-Conts and other GEM frame T-Conts. A case in which both kinds of traffic are carried within a 125-μs frame is referred to as **dual mode** usage.

Both the ATM flow and the GEM frame flow within a T-Cont may be multiplexed. One ATM T-Cont flow may contain several ATM virtual paths, which in turn may contain one or more ATM virtual circuits. The GEM frame flow may consist of several ports, each of which interconnects a particular pair of software or hardware entities, one at the OLT end and one at the ONU end.

In the incoming direction, the framing sublayer extracts the ATM cells, reads their addresses (actually, which VP they belong to) in the Alloc-ID component and then lets the TC adaptor filter them by VP number, after which the virtual path/virtual circuit identifier (VP/VCI) filter sends them to the right ATM client. For incoming GEM frames the same thing happens as the GEM TC adaptor filters them based on what the Alloc-ID block said their 12-bit port ID was, and sends them on for further filtering according to port IDs and port-type indicator (PTI) and then to the GEM frame client. In the outgoing direction, the process is reversed.

As shown in Figure 2.7*b*, the GPON transmission convergence (GTC) framing sublayer also includes a means for generating the PLOAM information (physical level operation and management) used for activation of the ONU in the first place, setting up a path between OAM instances in the OLT and ONU, and carrying the ranging information.

2.6.2   GPON Portrayed as Formats

In this section we discuss what the messages would look like to a technician with an oscilloscope somewhere between the OLT and ONU watching the bits go by. Figure 2.9 shows an overview of the syntax of the GPON **media access protocol**, which controls how addressed data is sent downstream, and later, how interleaved bursts of data from various ONUs are sent upstream. The interleaving is specified by the **upstream bandwidth map (USBW map)** portion of the **downstream physical level control block (PCBd)** that each ONU will have received from the OLT. The figure shows that it is permissible for a single ONU to have more than one T-Cont in the same frame, and these may be, but are not required to be, in immediate succession.

There are actually five different T-Cont types that define five different *classes of service*, as discussed in detail in [G.983.4, Section 8.3.5.10.2] and [Angelopoulos]. Traffic of different types are queued separately. The fact that classes of service are defined in detail is considered by its advocates to be one advantage of GPON over EPON.

We first treat the formats for the downstream case, and list what each piece does. The downstream bit stream is shown in Figure 2.10 as a series of exploded views of headers, subheaders, and so forth, plus the payload sections. The corresponding upstream flow is shown in Figure 2.11.

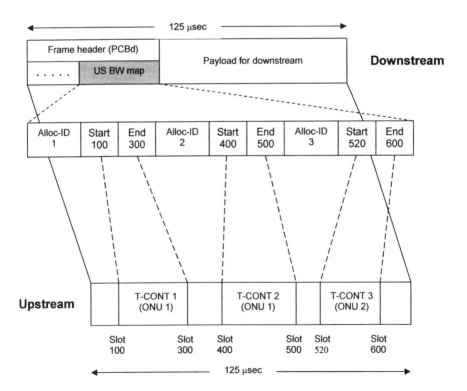

FIGURE 2.9 Summary of how in GPON the OLT sends a payload downstream to the various ONUs, and how these ONUs follow the OLT's instructions in interleaving their transmissions upstream.

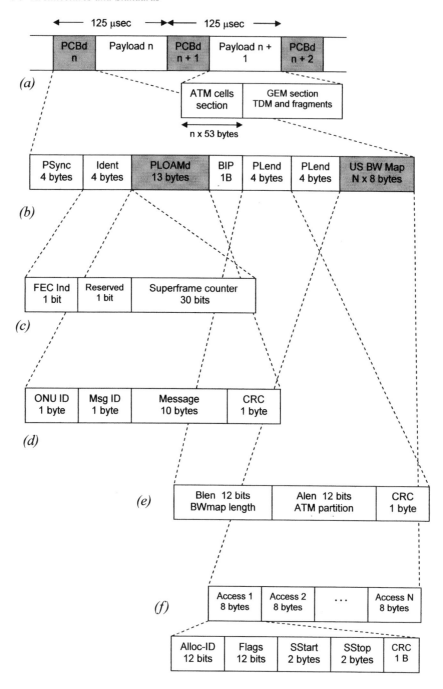

FIGURE 2.10 Successively exploded views of the downstream (OLT to ONU) GPON message syntax. The quantity N is the total number of addressable ONUs: (*a*) frames, (*b*) physical control block downstream, (*c*) ident, (*d*) downstream PLOAM, (*e*) payload length downstream, and (*f*) upstream bandwidth map.

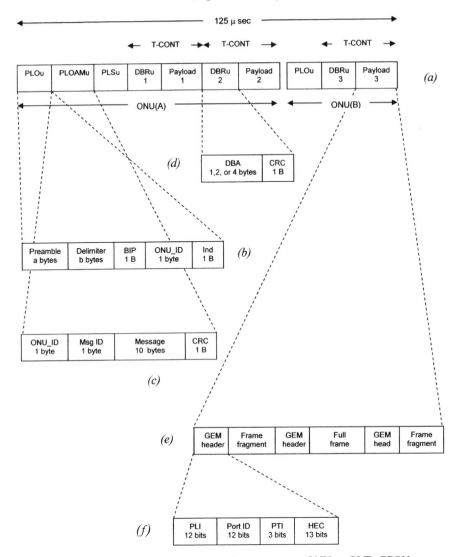

FIGURE 2.11 Successive exploded views of the upstream (ONU to OLT) GPON syntax: (*a*) Frames, (*b*) physical level overhead upstream, (*c*) upstream PLOAM, (*d*) dynamic bandwidth report upstream, (*e*) payload consisting of GEM frames and a fragment, and (*f*) GEM header.

As Figure 2.10*a* shows, each successive 125-μs frame consists of a downstream physical level control block (PCBd) of 4 bytes for downstream communication, and a payload consisting of:

- Some number of ATM cells at 53 bytes per cell.
- A GEM section carrying either plesiochronous data such as SONET and/or framed (packetized) data such as Ethernet packet fragments. "Plesiochronous" means that a transmitter–receiver pair are synchronized to each other but not necessarily to the rest of the network.

As Figure 2.10*b* shows, the PCBd is in turn composed of:

- The physical sync (Psync), an unvarying field of 32 bits as a marker for the ONU to use to find the beginning of the frame.
- The Ident field of 32 bits, shown in Figure 2.10*c*. Its purpose is to allow the option that a number of GPON frames can be considered as a larger group called a *superframe* by such higher-level things as encryption or applications using long packets extending over a succession of 125-μs frames. This feature also allows conveying low-rate synchronous reference signals to any higher-level function that might in the future need it. The first bit in this field to indicate whether *forward error correction* (*FEC*) is in use on this downstream link, the second is reserved, and the remaining 30 act as a counter that counts off frame by frame within the superframe, and if there are more than 2^{30} of them, the 30 bits simply wrap, restarting at zero.
- The physical layer OAM *downstream* (PLOAMd) field is the one that carries most of the downstream messages that support control functions that live in the control plane of Figure 2.2. The PLOAMd format is shown at Figure 2.10*d* and is seen to consist of 1 byte identifying the recipient ONU, 1 byte of message ID, a 1-byte field for a cyclic redundancy check of the entire frame, plus space for 10 bytes of PLOAM message. These messages may include the following, among other things:
  - The upstream *equalization delay* that the OLT wants the ONU to use as a result of the ranging calculation the OLT has made. Ranging was discussed in Section 2.5.4.
  - Asks if the ONU's *serial number* is what the OLT thinks it is.
  - Assigns a *temporary ONU ID* number to this equipment serial number.
  - Asks the ONU to *shut up and reset* itself.
  - *Disables* the ONU.
  - Tells the ONU which data flows to *encrypt.*
  - *Challenges* the ONU to give its password.
  - Asks the ONU to *generate a new encryption key* and send it to the OLT.
  - Commands the ONU to *change the transmitted optical power level* (important, because of the limited dynamic range of some photodetectors).
- The *bit interleaved parity* (*BIP*) is a 1-byte field that contains the parity of all the bytes transmitted since the last BIP, allowing the ONU to monitor the bit error rate.
- (*PLend*), the downstream *payload length* field, is not only protected with its own cyclic redundancy check (CRC) bits but also is transmitted twice. It is PLend (Fig. 2.10*e*) that specifies the location of the partition between the ATM part of the payload and the GEM part, as was shown in Figure 2.10*a*. It also gives the length of the field that immediately follows PLend, namely the upstream bandwidth map (USBW map).

- The USBW map (Fig. 2.10*f*) is where the OLT broadcasts to all the ONUs and gives each its marching orders about upstream transmission. It consists of some arbitrary number *N* of 8-byte *access allocation* structures, *N* being the total number of instances of an upstream T-Cont. *N* is therefore at least the number of ONUs on the network. The subfields of each allocation structure are

  - *Allocation ID*, which acts as the address, namely which particular ONU's T-Cont is being given upstream time for transmission.
  - *Flags* field, 12 bits that give the ONU a list of things it wants it to do, such as whether it should use CRC upstream, readjust power, include an upstream PLOAM (for OAM purposes), and other things.
  - *Start* and *stop fields*, which tell the ONU the start and end time of its upstream bandwidth allocation (its T-Cont).
  - Still another *CRC* field to protect just the individual upstream access allocation.

Figure 2.11 gives for the upstream case the same exploded view we have just discussed for the downstream direction. Remember from Figure 2.9 that each successive 125-μs upstream frame consists of a succession of T-Cont transmissions from a number of ONUs, separated by the small guard time interval between them. Now, Figure 2.11*a* shows two successive T-Conts, each with its upstream dynamic bandwidth report (DBRu) emanating from ONU (A), followed by the guard time, followed by one T-Cont issued by ONU (B).

The different pieces of the header, shown preceding the T-Conts in Figure 2.11*a*, are:

- The **physical level overhead upstream (PLOu)** (Fig. 2.11*b*), which is sent every time an ONU takes over the PON medium from another ONU. It consists of:

  - Several bytes of *preamble* and *delimiter*.
  - The same kind of *bit interleaved parity* byte that was seen in the downstream direction, namely the bit-interleaved parity of all bytes transmitted upstream since the last BIP from this ONU.
  - A 1-byte unique identifier for that particular ONU.
  - An 8-bit *indicator field* that tells the OLT what priority of waiting traffic the ONU is holding next in its buffers.

- The *upstream* version of the same **physical level OAM (PLOAMu)**, shown at Figure 2.11*c*, that we saw before in the downstream direction, with the very same allocation of the 13 bytes to carry control plane related information. For the upstream case the 10 bytes of PLOAM message include the following:

  - This ONU's *serial number* for the OLT to use in doing the ranging.
  - The ONU's *password*.

- The *dying gasp* to notify the OLT that it will do a normal power-off. This prevents the OLT from thinking that the ONU has died in its bed (failed from a malfunction) rather than deliberately committed suicide.
- An *encryption key* selected by the ONU.
- A count of the *number of errors detected* during a BIP interval.
- An *acknowledgment* of reception of a downstream message.
- The *power leveling sequence upstream (PLSu)* field, a 120-byte field reserved for the response to the downstream request, described under the PCBd (the downstream physical level control block), that asks for a change of transmitted optical power. The details are vendor specific.
- The dynamic bandwidth report upstream (DBRu), shown at Figure 2.11*d*, one for each T-Cont, containing the dynamic bandwidth allocation field, protected by its own adjacent cyclic redundancy check. This field may be optionally used to report to the OLT how many T-Conts are waiting in the buffer.
- The payload, as shown at Figure 2.11*e* for the GEM case, not the ATM cell case, consisting of a sequence of GEM headers plus fragments of the user-level frames.
  - The *GEM header* (Fig. 2.11*f*) consists of the *payload indicator (PLI)*, giving the number of bytes of the fragment to follow, the port ID, stating which of the 4096 ports on the network sent the traffic, the *payload-type identifier (PTI)*, telling the OLT something about local congestion, the *header error correction (HEC)* field, providing error detection and correction of the GEM header, and finally (!) the payload frame fragment.

2.6.3 GPON Portrayed as Sequences of Events

The GPON standards document [G.984.3] presents a long series of tables listing the various states that both OLT and ONU can occupy during transient activation, traffic flow, ranging, dynamic bandwidth allocation, and the many failure and recovery states that have to be considered. Perhaps a good flavor of what is going on can be gotten from Figure 2.12 from that document, which shows a possible event sequence at the ONU. This sequence includes ranging but does not include dynamic bandwidth allocation. The figure explains things in terms of the exchange of messages that use the bit and byte syntax just enumerated, but expressed in terms of "verbs rather than nouns." In other words, the present reader must go to the detailed documentation to find out which bits and which bytes in the explanation just given must be set at what values in order to map what has just been said about formats into the transitions between states in Figure 2.12. Nevertheless, it is instructive to follow what happens using this diagram.

There are only eight states for the ONU, of which two (O3 and O6) have two substates and another (O4) has three. The diagram in Figure 2.12 covers all relevant events from the initial power-on of the ONU (O1) to the steady-state transfer of data downstream (between states O1 and O2) or upstream (O6)—after all the ONUs have

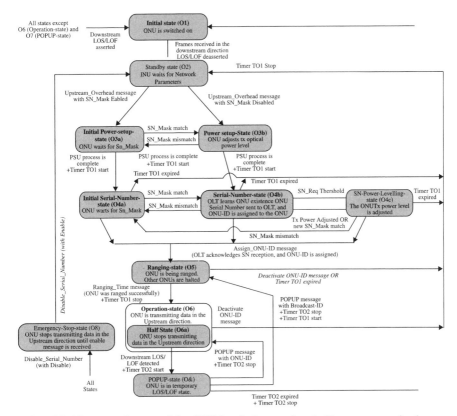

FIGURE 2.12 State diagram of the GPON optical network unit. Frames are received upon exit from state O1 and transmitted when in state O6. LOS = loss of signal, LOF = loss of frame, either at power-on time (before O1) or as a temporary condition ("POPUP state" O7) [G.984.3].

been contacted and have been synchronized together (ranged) so that their upstream transmissions will arrive at the OLT without overlap.

For GPONs, the semantic content of the dynamic bandwidth allocation, ranging, and protection switching functions are essentially identical to those already discussed with BPONs in Sections 2.5.2, 2.5.4, 2.5.5, and 2.5.6, respectively.

2.6.4 GPON Encryption

The documentation for GPON explicitly includes an encryption method [G.984.3]. It uses the Advanced Encryption Standard [AES] promulgated by the National Institute of Standards and Technology (NIST) as a successor to the 30-year-old Digital Encryption Standard (DES).

Only the user information is encrypted, not the many control fields that have just been enumerated. For ATM traffic, this means that each 48 bytes of each cell are exclusive-ORed with the output of the AES key generator for encryption, and since AES is a symmetrical crypto system, the same X-OR operation at the receiver recovers the cleartext. Each GEM fragment payload is similarly encrypted and decrypted.

There are a number of complexities that lead to key generator resets and even to the discard of a few extra bits due to the lack of natural time identity between key stream generator and GPON timing boundaries. For example, the counter of the AES key generator starts at the beginning of the frame, but the actual encryption aligns with the data payload. Special measures are taken to ensure that the same AES key generator starting condition is never used more than once. The crypto key may be constant over some period of time, but the key stream will not repeat.

The key is created by the ONU when it receives a *key_request* message in a PLOAM from the ONT. The ONU responds by creating, storing, and then sending the key it has generated to the ONT over a succession of PLOAM messages. It does so three times to ensure accuracy.

■ 2.7 IEEE ETHERNET PASSIVE OPTICAL NETWORK (EPON) STANDARD 802.3AH

2.7.1 EPON Portrayed as Layers

Figure 2.13 gives a layer view of the EPON architecture [802.3ah]. As before, the subscriber's ONU has one instance of all the communication-level layers,

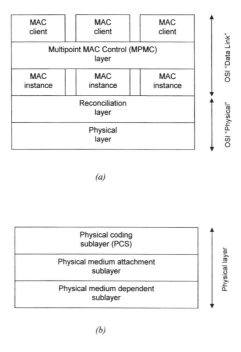

FIGURE 2.13 Protocol stack for data flow in an EPON. The stack for control (initiating, terminating, and recovery of traffic flows) is not shown: (*a*) Principal layers and (*b*) physical layer sublayers.

whereas at the head-end OLT there are a number of instances of most layers equal to the number of active ONUs. Similarly, the first two OSI layers of Figure 2.2 have a correspondence with those of EPON. The EPON layers up through the reconciliation layer correspond to the OSI physical layer, and those from the MAC up to the MAC client layers of EPON correspond roughly to the OSI data link layer.

The physical (PHY) layer is seen to be formally subdivided into

- The **physical-medium-dependent (PMD)** sublayer, embodying essentially all the photonics—the laser, its driver, the photodetector and its associated electronic amplification, and the connections to the actual outplant fiber.
- The **physical-medium-attachment (PMA)** sublayer, which converts from a serial bit stream below in the diagram to a parallel one upward while adding a simple (per byte) form of error control code.
- The **physical coding sublayer (PCS)**, which turns the laser completely off between transmissions, and also optionally provides an extra level of forward error correction.

The EPON Physical-Medium-Dependent Sublayer

One should note a difference in nomenclature—for BPON and GPON the term *physical-medium-dependent layer* meant the same kind of table of optoelectronic specifications as that of only the *sublayer* of the same name in EPON parlance. Here is the EPON version:

- Bit rates: 1.25 Gb/s both upstream and downstream (within 1 part in 10^4)
- Wavelengths: 1260 to 1360 nm up, 1480 to 1580 down
- Traffic type: digital only
- Fiber splits: 16 typical; more with forward error correction (FEC)
- Worst-case attenuation permitted in the optical distribution network (ODN) separating OLT and ONU: Upstream at 1310 nm, 5 to 20 or 10 to 24 dB for ODN length of 10 or 20 km, respectively; downstream at 1550 nm, 5 to 19.5 or 10 to 23.5 dB for 10 or 20 km, respectively
- Launch power range downstream: -3 to $+2$ dBm for 10 km ODN length or $+2$ to $+7$ for 20 km
- Launch power range upstream: -1 to $+4$ dBm for both 10- and 20-km ODN length
- Worst-case receiver sensitivity/receiver overload: $-24/+4$ dBm and $-27/+4$ downstream for 10 and 20 km, respectively; upstream $-24/+2$ and $-24/+7$ for 10 and 20 km, respectively
- Maximum bit error rate: 10^{-12}
- Laser extinction ratio: >6 dB
- Laser linewidth for 1490 nm downstream: 0.6 and 0.3 nm for 10 and 20 km, respectively; for 1310 upstream, 3.5 and 2.5 nm for 10 and 20 km, respectively.

The EPON Physical Medium Attachment (PMA) Sublayer

This sublayer is the place where the well-known Franaszek–Widmer 8b10b line code is applied to the serial bit stream to convert it from 8 bits per byte in the layer above to 10 bits for those below, in the process providing considerable error protection while at the same time preserving DC balance, even when there is a long string of 0's or 1's to be encoded. In the process, the raw 1.25-Gb/s bit rate seen out in the ODN itself gets converted to 1.00 Gb/s, as seen within the higher layers of both the OLT and the ONUs.

The EPON Physical Coding Sublayer (PCS)

The laser turn-off function provided here is different from setting the laser to send logical 0's. As we shall see in Chapter 3, and as indicated by the "extinction ratio" entry in the list just given, binary 0's are encoded not by turning the laser completely off but by reducing the output by numbers like 6 dB (factor of 4) for EPON and 10 dB (factor of 10) for BPONs and GPONs, as listed earlier in Sections 2.5.1 and 2.6.1. However, to keep the laser from emitting noise in the form of **spontaneous emission** when it is some other ONU's turn to talk, the drive current of each of the other lasers must be driven to zero. Binary 0's cannot be encoded as zero laser drive because, when the next 1 arrives, it takes a few nanoseconds for the laser to start emitting the desired steady coherent light (**stimulated emission**), as we shall see. The PCS layer of EPON detects the presence of bits from the higher layers and buffers these long enough for the laser to get started and spit out a few idle bits. The converse operation takes place at the end of this ONU's turn to talk.

An optional heavy error control takes the form of a Reed–Solomon block code imposed at the sending node and decoded at the receiving node in such a way that, even if the bit error rate gets as high as 10^{-4} when the bit stream arrives, it has been reduced to 10^{-12} before being passed upstairs to the higher layers. At the transmitter, the process takes each block of 239 bytes and adds 16 data-dependent parity bytes to form a 255-byte block that the receiver then converts back to a 239-byte block.

The EPON Reconciliation Sublayer

This is just a fancy name for the function of routing the packets within a node that may have multiple addresses within that node, in particular the OLT. In the ONU, there is only one instance of the higher layers, so in this case the reconciliation sublayer is degenerate (no decoding or encoding of internal addresses required).

The EPON Media Access Control (MAC)

The MAC layer is clearly where all the interesting things happen. It is from this layer that the OLT issues grants to the various ONUs, allowing them to transmit data for a certain length of time before relinquishing the shared fiber medium to some other ONU. It is also the place in the layer structure at which the ONU receiving one or more grants acts upon this information.

The Multipoint MAC Control (MPMC) Layer

Whereas there can be multiple instances of the MAC layer (and obviously of the client layer at the top of the stack), there is only one instance of the **MPMC** layer. The multiple access protocol executed by the MPMC layer arbitrates between the many ONUs by granting a mutually nonoverlapping upstream transmission window to each (analogous to the T-Cont of APONs).

When the network is first activated, there is an automatic discovery sequence culminating in binding an LLID (logical link ID) to one of the MAC instances in the OLT and its partner single instance MAC in a particular ONU. Thus the MACs are uniquely identified by their LLIDs.

Thereafter, using ranging, the MPMC layer handles time-interleaved downstream and upstream transmissions across the ODN, acting as a client for the multiple MAC instances at the protocol level below and serving multiple client instances at the higher levels.

2.7.2   EPON Portrayed as Formats

Figure 2.14*a* shows a typical EPON frame. It is essentially the same as a standard gigabit Ethernet frame [Seifert, Tanenbaum] but with the first 8 bytes changed to reflect the PON topology. The total length can be anything between 72 and 1974 bytes because the data field can be anything between 46 and 1500 bytes. Thus, at the short end, an EPON frame is not that different in size from a 53-byte ATM cell, while at the long end the size is comparable to the large block sizes used in virtual memory storage [Hennessy].

For data transfer (not control) the data frame consists of the following fields:

- An 8-byte preamble/SFD (start of frame delimiter) field. As shown at Figure 2.14*b*, this field is in turn subdivided into:
  - One byte of start of LLID delimiter (SLD)
  - Two bytes of fixed bit patterns to define the SFD function
  - One LLID byte, except that the first bit is the *mode* bit—0 if the issuing node acts as an OLT and one if as an ONU
  - A second full LLID byte; with the first LLID one can distinguish up to 128 source or destination MAC instances, and with the second another 256-fold multiplicity
  - An 8-bit cyclic redundancy check of the LLID fields
- Six-byte *source and destination addresses* (Fig. 2.14*d*). Single-station destination addresses have 0 for the highest order bit, while group addresses (multicast) use a 1, including full broadcast to *all* ONUs, in which case the field is all 1's. This leaves 47 bits. One of the consequences of the very large Ethernet address space (some $7 \times 10^{13} - 1$ resolvable addresses) is that every Ethernet card or unit shipped can be manufactured with a unique identifier— each separate manufacturer is given a block of the 23 higher order bits by a central authority, and he then uses the remaining 24 bits to burn into the electronics of each instance of his product line output the address that that product

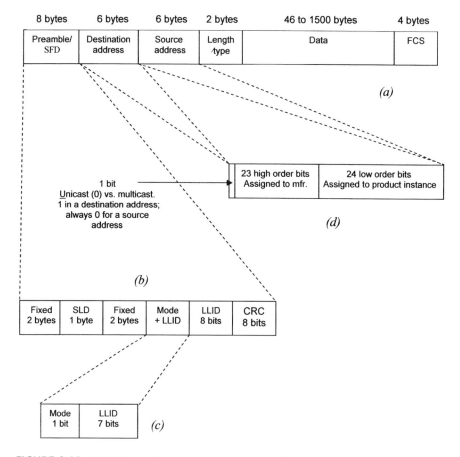

FIGURE 2.14 EPON frame format for data transfer: (*a*) Overall data format, (*b*) and (*c*) the 8 bytes of preamble decomposed, and (*d*) reserved and available portions of the two address fields.

instance permanently uses for outgoing communications and that others use to send it a frame. This is to be contrasted with the BPON and GPON addressing conventions in which BPON allows 8 bits, giving an address space of 256 within each PON (in the MSG_PON_ID byte within the PLOAM) as does the GPON architecture (in the message ID byte in the PLOAM).

- The length/type field (2 bytes) can be used to tell how many bytes constitute the data field, which immediately follows, or can alternatively be used to say what type of OAM frame this one is.

- The data field has a maximum length of 1500 bytes, set by buffer delay considerations, but it also has a minimum length of 46 bytes, inherited from earlier forms of Ethernet in which a transmitter might otherwise think it had successfully completed its transmission only to find out from the far end that the far end had inadvertently started its own transmisson before realizing that it should be receiving and so had immediately ceased data transmission.

- The final frame check sequence, 4 bytes long, protects the data field by detecting errors in the received version. The checksum does not protect fields earlier than the data field.

As was mentioned, at the ONU there is only one instance of the MAC and one of its client upper layers. Clearly, at the OLT there are as many instances of both of these as there are active ONUs. These instances are realized concurrently, so as to permit broadcast or multicast. Also the MPMC layer must be implemented in a multithreaded way in order to support broadcast situations. The logical link ID is the field used to bind a MAC instance to its designated physical ONU. It is carried in the first field of a data frame, as shown in Figures 2.14*b* and 2.14*c*. This field is what used to be the 8 bytes occupied by the preamble and start-of-frame delimiter in gigabit Ethernet and earlier versions of Ethernet. The LLIDs are assigned to the different ONUs when they are initially registered with the OLT.

When a frame is used for control rather than data transmission purposes, the format is quite a bit different—specifically, the first 8 bytes are missing and the data field length is frozen at 40 bytes. The two varieties of these MPMC protocol data units are shown in Figure 2.15, which carry bandwidth capacity

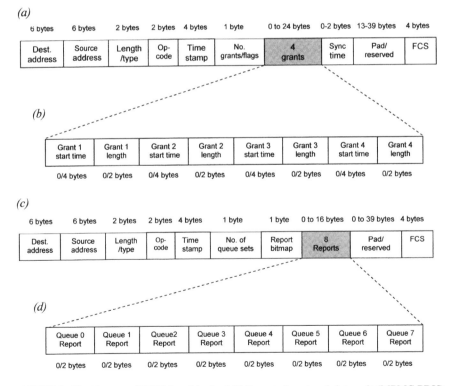

FIGURE 2.15   Format of EPON multipoint MAC control protocol data unit (MPMC PDU): (*a*) For GATE downstream from OLT, (*b*) four grants per GATE PDU, (*c*) for REPORT upstream from ONU, and (*d*) eight reports per REPORT PDU.

grants out to the ONUs, and reports back to the OLT. We discuss the process shortly.

2.7.3 EPON Portrayed as Sequences of Events

As we have seen, in the EPON architecture, it is the MPMC layer that is in charge of controlling the flow of data transfer frames (Fig. 2.14). It does this by exchanging MPMC PDUs (protocol data units) between the MPMC instance in the OLT and the instances in the many ONUs. These control frames take priority over the data frames to and from the clients in Figure 2.13*a*. The data frames may be delayed or even discarded in the process.

The OLT sends data to a particular ONU by broadcasting suitably addressed data frames, and an ONU will send to the OLT using the same frame format. The process of interleaving upstream frames in a nonoverlapping way is much the same as was discussed earlier for the BPON and GPON cases, as we shall now see.

Dynamic Bandwidth Allocation

In addition to knowing when it can start transmission from the ranging results, each ONU is also given a number of grants by the OLT, the number of frames it many send and what their lengths are to be. The GATE protocol data unit (Fig. 2.15*a*), addressed to it by the OLT, tells it these things. The OLT has based its allocation of number of frames to the various ONUs by listening to the REPORT PDU that it received earlier from each ONU (Fig. 2.15*c*), telling the OLT—in one 8-bit queue report each—the length of each of its separately prioritized queues. The Ethernet architecture allows each ONU to have up to eight of these queues at different priority levels. The REPORT bitmap field of 8 bits says, using binary 1 or 0, respectively, whether or not for each possible queue there even exists such a queue.

Other fields are as follows:

- Source and destination addresses and length/type are as discussed before for data frames.
- The opcode specifies the type of MPMC protocol data unit that this is, for example, GATE or REPORT.
- The time stamp field which supports the ranging function, which we discuss in the next section.
- The number of grants/flags field of GATE, which, among other things tells how many grants will follow. For REPORT, this byte states the number of queue sets where data awaits transmission upstream.
- In REPORT there then follows the 1-byte report bitmap indicating which of these upstream queues are even present.
- There then follow four grants in the case of GATE or eight queue reports for REPORT. The grants tell the ONU the start time and length of each granted transmission interval. The queue reports give the length of each permitted queue in the ONU. For both REPORT and GATE, respectively, the grant

field or the queue report field, respectively, is actually absent when that OLT grant is not to be issued or that ONU queue is absent.

• The PAD/reserved field is all zeros for both GATE and REPORT. It is used for other types of PDUs not covered here.

As Figure 2.15*d* indicates, there can be up to eight different differently prioritized queues of traffic at any ONU, and these can be used to define various classes of service, analogous to the five different T-Cont types in BPON and GPON (collectively known as APON). In the EPON case, the exact restrictions on each of the eight classes are still being defined [Angelopoulos].

Ranging

Note the provision of a timestamp field in Figures 2.15*a* and 2.15*b*. It conveys the content of a local 32-bit register that keeps time at the ONT or ONU. Each ONU knows when it is allowed to send because the OLT has told it what time offset to use by means of the ranging protocol. This protocol operates in exactly the same way as the BPON and GPON ranging protocols described earlier, apart from differences in syntax and in the amount of guard time that is allowed.

■ 2.8 COMPARISON OF ATM-BASED AND ETHERNET-BASED PONs

Now that we have worked through some of the layer and format details of BPON, GPON, and EPON standards, it is interesting to summarize what each accomplishes in a comparative way. Figure 2.16 gives some highlights.

Several things about Figure 2.16 deserve further comment. Some of the differences reflect the cultural differences of the communities that created the standards. We have already mentioned the telcos' inherited obsession with voice traffic and the consequent emphasis on cells rather than packets, an emphasis that is dissipating year by year.

The carrier community is traditionally much more obsessed with data integrity during transmission than is the DP community, which has relied heavily on retransmissions. This is evidenced not only by such things as the traditional protection switching of SONET/SDHs versus the "shoot-and-hope" nature of Internet IP, but in the present context also by GPON's inclusion of encryption, plus the extensive use in both BPON and GPON of limited-span CRCs. This is to be contrasted with EPON's use of a single 4-byte frame check sequence protecting the entire data field plus the two address fields the header, plus the CRC protecting the 8-byte preamble. Also, reflecting data integrity considerations is BPON and GPON's inclusion of five separately queued well-defined T-Cont traffic classes, as contrasted with the eight queues of EPON, whose restrictions are, to date, only partly defined.

On the other hand, the clever use of a huge 48-bit address space in EPON has facilitated cranking out millions of components that will always be self-identifying no matter how the Ethernet market expands. The BPON and GPON address spaces

Parameter	BPON (G.983)	GPON (G.984)	EPON (802.3ah)
Bit rates down	155 and 622 Mb/s	1.2 or 2.4 Gb/s	1.25 Gb/s
Bit rates up	155 Mb/s	155, 622 MB/s; 1.2 or 2.4 Gb/s	1.25 Gb/s
Packet capability	Cells only	Fragmentation every 125 μs	Native, but fragmentation allowed
Analog video	Specified	Not specified	Not specified
Protection switching	Specified	Many options	Not specified
Encryption	Not specified	Uses AES	Not specified
P2P also?	Not specified	Specified	Specified
No. ONUs	Up to 32, limited by attenuation	Up to 64, limited by attenuation	16 typical if no FEC; limited by attenuation
Error protection	Certain control fields + polynomial code for each cell	Certain control fields. No line code	Preamble + 8b10b line code + Reed–Solomon FEC
Address space	8 bits	8 bits	48 bits
Class of service	5 T–Cont types	Same	8 queues

FIGURE 2.16   Feature comparison of BPON, GPON, and EPON.

span not much more than a single PON. Perhaps the most striking difference, in this modern world of IP packets, the Web and ubiquitous small, cheap laptops and desktops, is the complexity of the APON family (BPONs and GPONs) compared to EPONs. (Compare the steady-state syntax of Fig. 2.10, showing 20 separate forms of subfield, with that of Fig. 2.15, showing 10.) Part of this complexity is a consequence of the telcos' relatively strong obsession with data integrity, but much is due simply to the tyranny of the 125-μs framing, inherited from T-carrier to BISDN to PONs. Forcing the 125-μs boundaries is what leads to the elaborate fragmentation and reassembly described in connection with Figure 2.8. With EPONs, the packet can be almost completely user specified, and fragmentation is not obligatory.

The integrity-driven complexity of APONs compared to EPONs is reminiscent of the history of IBM's late-lamented System Network Architecture [Green-1], as compared to "shoot-and-hope" IP, which now dominates all telecommunications, shortly to include even voice.

While BPON and GPON are dominant in today's implementations in the United States, EPONs are widely used elsewhere, as we shall see in Chapter 5. It is this author's prediction that EPONs and their descendants are likely to become the norm everywhere, for four reasons. First is EPON's relative simplicity and therefore lower first cost, maintenance cost, and ease of system design. Second, Ethernet parts have enjoyed a 25-year learning curve of cost reduction, and this cost curve is endemic in the computer world but rarely seen in the telco world. Over the 25-year period, Ethernet parts have been assembled into a variety of networks that have proved to be as resilient as most users seem to feel they will ever need. Third, IP packets flow over an EPON natively, rather than requiring protocol conversion, of which fragmentation/reassembly is just a part. Fourth, it will be easy to make the change from APONs to EPONs—only the ends of the PON need to change, while the interfaces to the backhaul environment at the CO and that to each user at the ONU can largely remain the same (Fig. 2.1). In other words, while changing the PON from an APON to an EPON may have some effects on the OAM functions, it need not affect the steady-traffic interfaces at each end.

As time goes on and today's ILECs follow their success with wireless by broadening businesses into new areas such as VoIP, IP service provision, video (either analog or IP based), and so forth, none of it based on the classic black telephones, then the heritage of ATM cells and 125-μs framing are increasingly likely to be seen as an expensive luxury from the past, eventually achieving only a lingering archeological significance.

Note that in Figure 2.16 the point-to-point (P2P) option is mentioned. This option is certainly being implemented in some communities, but it is very much a minority component of fiber to the home because of its expense in extra fiber counts along the right-of-way and the larger number of optoelectronic ports at the head end. The reader who is interested in the technical details of P2P architecture and standards can consult either the GPON standard [G.984.1, G984.2, G.984.3], the EPON standard [802.3ah], or any of the many excellent references on P2P Ethernet, for example [Seifert].

■ 2.9 AN EXAMPLE OF ARCHITECTURE VS. IMPLEMENTATION

The architecture standards documents we have just summarized, G.983, G.984, 802.3ah, are intended to present an unambiguous specification of what is implemented, but not how it is implemented. For example, the documents never say what is hardware and what is software, what is hard-wired, burned into a chip, or loaded at system initialization time. About all the standards do is to distinguish between what syntax and semantic content is fixed and what is user settable. The standards are unambiguous, and this is in principle all that an implementer needs in order to manufacture and activate a product that will interact with others on the PON obeying the same standard, give or take a few implementation-specific parameter settings.

The actual implementations involve many issues not contemplated in the standards—integration, power supply, physical layout, packaging and so forth. A very instructive example of how the architecture appears when actually implemented is given in Figure 2.17 [Seifert] for the case of a Gigabit Ethernet product, essentially what would be implemented in a single EPON ONU. For a PON, the LLC and MAC

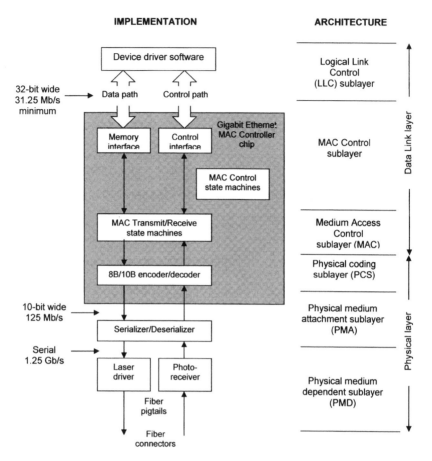

FIGURE 2.17 Implementation vs. architecture—an example equivalent to an EPON ONU [from Seifert, *Gigabit Ethernet: Technology and Application of High-Speed LANs*, © 2004, 1998 Pearson Education Inc. Reprinted by permission of Pearson Education, Inc. Publishing as Pearson Addison Wesley.]

sublayers would reflect the content of a single instance of multipoint MAC control and MAC client. The OSI layers of Figure 2.13 are indicated by the arrows at the far right, the various EPON layers by the list next door, and the hardware and software by the elements of the block diagram at the left. In this particular case, there are four chips [laser/driver, photoreceiver, serializer/deserializer (with line coding)/decoding, and a controller integrated circuit]. Figure 2.17 is seen to bear little resemblance to the many earlier figures in this chapter.

REFERENCES

[802.3ah] Part 3 of 802.3, amended, Media Access Control Parameters, Physical Layers, and Management Parameters for Subscriber Access Networks, IEEE Computer Society, 7 September 2004, clauses 60, 64, 65, and Annex 4A.

[AES] National Institute of Standards and Technology, *Advanced Encryption Standard*, Federal Info. Processing Standard 197, November 26, 2001.

[Angelopoulos] J. D. Angelopoulos, H.-C. Leligou, T. Argyriou, and S. Zontos, Efficient Transport of Packets with QoS in an FSAJ-aligned GPON, *IEEE Commun.*, Feb. 2004, pp. 92–98.

[ANSI/SCTE] American National Standards Institute, Society of Cable Television Engineers, Digital Cable Network Interface Standard,

[DePrycker] M. dePrycker, *Asynchronous Transfer Mode*, Ellis Horwood, Chichester, UK, 1991.

[FCC] Federal Communications Commission, Code of Federal Regulations, Title 47, Part 76.605, 2003.

[G.983.1] International Telecommunications Union, Broadband Optical Access Systems Based on Passive Optical Networks (PON), October, 1998.

[G.983.2] International Telecommunications Union, ONT Management and Control Interface Specification for ATM PON, April, 2000.

[G.983.3] International Telecommunications Union, A Broadband Optical Access System with Increased Service Capability by Wavelength Allocation, March, 2001.

[G.983.4] International Telecommunications Union, A Broadband Optical Access System with Increased Service Capability Using Dynamic Bandwidth Assignment, November, 2001.

[G.984.1] International Telecommunications Union, Gigabit-Capable Passive Optical Networks (GPON): General Characteristics, March, 2003.

[G.984.2] International Telecommunications Union, Gigabit-Capable Passive Optical Networks (GPON): Physical Media Dependent (PMD) Layer Specification, March, 2003.

[G.984.3] International Telecommunications Union, Gigabit-Capable Passive Optical Networks (GPON): Transmission Convergence Layer Specification, February, 2004.

[GR-909] Generic Criteria for Fiber in the Loop Systems, Telcordia Technologies, 2004.

[Green] P. E. Green, Jr., *Fiber Optic Networks*, Prentice Hall, Englewood Cliffs, NJ, 1991.

[Green-1] P. E. Green, Jr. (Ed.), *Computer Network Architectures and Protocols*, Plenum, New York, 1982.

[Hennessy] J. L. Hennessy and D. A. Patterson, *Computer Architecture—A Quantitative Approach*, Morgan Kaufmann, Palo Alto, 1990.

[Ireland] R. Ireland, The Emerging IP Revolution: A Wireline Example, 2005 presentation.

[Isenberg] D. S. Isenberg, The Rise of the Stupid Network, *Computer Telephony*, August, 1997, pp. 16–26, or www.hyperorg.com/misc/stupidnet.html.

[ITUI.732] International Telecommunications Union, Functional Characteristics of ATM Equipment, 2000.

[Lee] S.-M. Lee, et al., Dense WDM-PON Based on Wavelength-Locked Fabry-Perot Lasers, Paper JWA55, OFC/NFOEC Conference, March 9, 2005.

[Perkins] B. Perkins, The Art of Overlaying Video Services on a BPON, 2004 FTTH Conference, Paper T03.

[Poynton] C. Poynton, Digital Video and HDTV, Morgan Kaufmann, San Francisco, 2003.

[Ross] K. Ross, POTS to Pipes, *Network Technology Report*, February 9, 2005, www.broadbandpub.com.

[Seifert] R. Seifert, *Gigabit Ethernet*, Addison Wesley, Reading, MA, 1998.

[Tanenbaum] A. Tanenbaum, *Computer Networks*, 3rd ed., Prentice Hall, Englewood Cliffs, NJ, 1996.

a VOCABULARY QUIZ

Discuss not only what these terms abbreviate but also what they mean.

Adaptation sublayer	GEM	PC sublayer	SLM laser
All-optical network	GPON	PCBd	Spontaneous
APON	Grants	PCBu	emission
ATM	Guard time	Physical medium	Start of frame
BPON	Header	layer	delimiter
Bit-interleaved	HEC	Plesiochronous	State diagram
parity	Hitless switching	PLOAM	Stimulated
Cell	ITU	PMA sublayer	emission
Control plane	IEEE	PMD sublayer	Subheader
Cyclic redundancy	Laser control field	POP	Sublayers
check	Layers	Preamble	Syntax
Data plane	MAC	Protection	T-Cont
Delimiter	Media access	switching	Transmission
Dual mode	protocol	Protocol data unit	convergence
Dynamic bandwidth	Mixed mode	Protocol	layer
allocation	MLM laser	layering	Transmission
EFM	MPMC	Ranging	sublayer
Embedded OAM	OAM	Receiver control	USBW map
EPON	Optical distribution	field	Virtual circuit
Equalization delay	network	Reconciliation	Virtual path
Frame	OSI	sublayer	
FSAN	OSS	Security	
G-EPON	Overlay	Semantics	

CHAPTER 3

Base Technologies

In this chapter, we discuss the piece parts that make up the overall systems treated in the preceding chapter. Then in Chapter 4 we shall see how these systems are actually deployed in the field.

The photonic components business is highly evolved, both with respect to the functions performed and the reliability, diagnosis, and management of these elements. However, it has not completely evolved to the lowest possible cost. Forty years of preoccupation with very large transmission distances, many concurrent wavelengths, and aggressive bit rates per wavelength have certainly produced a huge backlog of solutions to pick from in creating FTTH systems. But in the access environment, unlike the interexchange and metro environments, cost often prevails over function. Thus, the principal technology challenge of FTTH has been to cost reduce the appropriate subset of historically available solutions, rather than to invent new ones.

Therefore, while the material of this chapter will be old hat to those familiar with the fiber-optics communications field up to now, the necessary emphasis on cost reduction leads to some interesting new wrinkles.

■ 3.1 OPTICAL FIBER BASICS

The conventional wisdom is that what is most important about optical fiber made of silica (silicon dioxide) is that it has such an incredibly broad passband. Just in one of several bands, fully 25,000 GHz of bandwidth (35 nm) are available. As shown in Figure 3.1, the **C-band**, roughly centered on the attenuation minimum at 1550 nm, is the one used in current FTTH deployments to carry the downstream video component of the triple-play services, the highest bit rates of which are expected to be no more than a mere 10 to 100 Gb/s per user in the time-interleaved fashion described in the preceding chapter. The **O-band** includes the 1310-nm wavelength of the upstream data, and the **S-band** the 1490-nm wavelength of the downstream

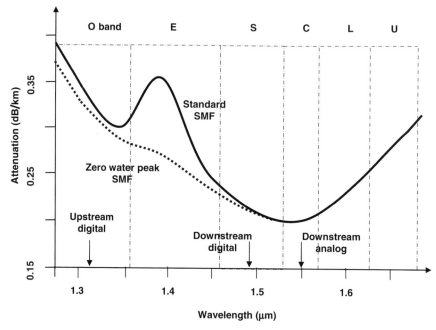

FIGURE 3.1 Attenuation of fiber as a function of wavelength. The dashed curve is for the zero water peak (ZWP) fiber, as standardized in [ITU G.652C].

data. Even with such broadband traffic, fiber, with its 100,000 GHz of exploitable bandwidth, provides a flagrant case of technological overkill.

But it is not just the bandwidth that is so significant about this transmission medium. The low attenuation (0.20 dB/km at 1.55 μm, as shown in the figure), the low cost (less than copper), the small physical size and weight of fiber cable bundles, the almost complete absence of any known aging mechanisms, imperviousness to lightning strokes, and the fact that no superior medium has been found since silica fiber was introduced in 1965—these all argue that an investment in connecting the world with fiber will be as permanent as we are ever likely to have. Every once in a while a minor challenge to the "future-proofness" of silica fiber shows up, as happened in the mid-1990s with fluoride fiber, with its astonishingly low attenuation of 0.001 dB/km, but somehow there always seems to be something else wrong with each alternative to silica, in this case the fact that the fluoride fiber proved to be soluble in water.

There are basically two types of fiber, **single-mode fiber (SMF)** and **multimode fiber (MMF)**. As Figure 3.2 shows, there are subsequently two forms of SMF, the conventional kind, shown at Figure 3.2*b*, and the very recently introduced **holey fiber**, or **hole-assisted fiber**, or **photonic bandgap fiber**, shown at Figure 3.2*c*. Holey fibers, which do confinement of the light by guiding between longitudinal holes rather than by index changes, have found a single use in FTTH systems to date, namely for cases in which extremely tight bend radii are required, as will be discussed in Section 4.10.

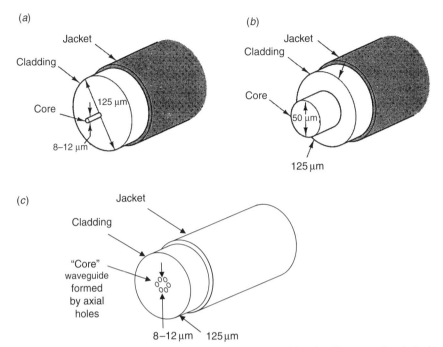

FIGURE 3.2   Cross sections of three forms of fiber: (*a*) multimode, (*b*) conventional single mode, and (*c*) hole-assisted single-mode fiber.

A propagation **mode** in a fiber is a distinct sideways standing-wave pattern that the fiber core will support. In any kind of single-mode fiber, the diameter of the **core** where the light is confined is so small that only one such mode can propagate, whereas with multimode, there are a number of them, each having its distinct propagation velocity. This produces significant multipath smearing or **modal dispersion**—the appearance at the receiver of a smear of many variously delayed replicas of the original transmission.

As Figures 3.2*a* and 3.2*b* show, the core is the central part of the cross section within which the propagating wave is trapped by internal reflection from the core-cladding boundary, where the cladding has a slightly lower index of refraction than the core. The cladding surrounding the core is typically 125 μm in diameter for both SMF and MMF and is in turn surrounded by a protective jacket. In the early days of optical fiber communication, there were very strong mechanical reasons to maximize the diameter of the core, in spite of the limitations on signaling speed and distance imposed by the modal dispersion.

A cross-sectional view of a few of the permitted propagation modes is shown in Figure 3.3. By making the core sufficiently small, one can suppress all but the lowest one, called HE_{11}, which is seen to look like a half-wave of a sinusoid fitted into the core diameter. For a given core diameter, all the higher modes would need to be at higher frequencies in order for any of them to fit.

Making the core large, as with MMF with diameters of 50 to 62.5 μm, makes splicing, connectorization, and cleaning much more convenient than the nightmare

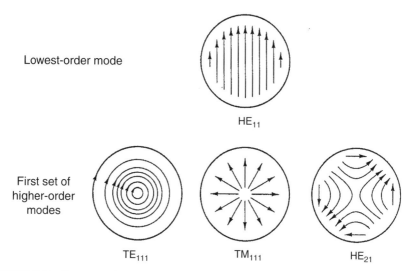

Lowest-order mode

HE_{11}

First set of higher-order modes

TE_{111} TM_{111} HE_{21}

FIGURE 3.3 Some permitted propagation modes within the fiber core.

of doing these things with single-mode fiber, whose diameter is only 8 μm, about one-sixth that of a human hair. Recently, **large mode field diameter SMF** has been developed that spreads the field strength more evenly across the core in order to minimize the maximum field strength anywhere. This is done not for ease of handling but to mitigate the nonlinear effects that we shall discuss in Sections 3.2.3 through 3.2.5.

Today there is still a great deal of multimode fiber around, particularly within user premises, partly because for a long time it was so much cheaper to manufacture than single mode. This is no longer true since huge economies of scale and improvements in single-mode manufacturing methodologies have occurred, as have easier ways of splicing and connectorizing single-mode fiber, even bundles or ribbons. Today, MMF actually costs more than SMF because of the high cost of germaniam with which the cores are doped. Still, there are many low-bit-rate short-distance applications, for example, within the backplane of a computer or a piece of telephone equipment, where ease of fiber handling is paramount, and multimode still finds many such applications. Recently, significant multimode fiber cost reductions have been possible using plastic fiber over short distances, for example, within a single automobile, boat, or aircraft, where distances are so short that the much higher bulk material attenuation of plastic compared to glass is of little concern.

For most single-mode fiber great pains are taken to see that all the light energy is confined to the core as the HE_{11} mode, as was shown in Figure 3.3. However, there are circumstances in which it is desired to send some of the light out into the cladding, notably, with couplers and splitters, as will be discussed in Section 3.4. All that is required to do this is to narrow the fiber core even thinner than the normal 8 μm. The fiber is still single moded, but much of the energy is squeezed out into **evanescent** propagation outside the core and can transport no energy because the E-field and the H-field are 90° out of phase. However, in Section 3.4 we shall see that the evanescent wave modes from two closely adjacent fiber cores of very small diameter can be made to exchange actual power.

▪ 3.2 IMPAIRMENTS

As people began to use fiber for higher and higher bit rates with more subscribers served per fiber, and therefore greater transmit power and modulation bandwidth, various impairments began to show up [Green]. These include:

- **Loss**, which we have already discussed.
- **Chromatic dispersion**, the wavelength dependence of propagation velocity, which produces a time smearing of different portions of any wideband signal.
- **Scattering** of the light in the core into wasteful or even harmful directions. In decreasing order of importance for today's FTTH systems, these include:
  - **Rayleigh scattering**
  - **Stimulated Brillouin** backward **scatter (SBS)**
  - **Stimulated Raman** forward **scatter (SRS)**
- And various intermod effects:
  - **Self-phase modulation**
  - **Cross-phase modulation**
  - **Four-wave mixing**

The first three of these—loss, chromatic dispersion, and Rayleigh scattering—occur independently of signal power level, whereas all the others [fiber-optics.info] are a consequence of an increase of index of refraction with power. That is to say, the medium is slightly *nonlinear*. Specifically, the index of silica is 1.47000 at zero optical power in the core and increases linearly to 1.47006 at 130 mW, as given by the equation

$$n = n_0 + n_2 P / A_{\text{eff}}$$

where n_0 is the index of the core at low optical power levels, n_2 is $2.35 \times 10^{-20} \, \text{m}^2/\text{W}$, P is the optical power level in watts, and A_{eff} is the number of square meters of effective area of the fiber core. One might wonder how a change in index in the sixth significant figure could possibly matter to anyone. However, when the diameter of the fiber core is only 8 μm, it does not take much power P to produce a significant n_2 multiplier. Also, think of how many wavelengths of propagation length are contained in 10 or 20 km of fiber.

With some of the power levels involved in sending analog TV at 1550 nm in the systems we discussed in the last chapter, nonlinear effects can emerge as "show-stopper" impairments. Directly modulated lasers radiate typically 0 to 10 dBm (1 to 10 mW), but if one uses **erbium-doped fiber amplifiers (EDFAs)** to achieve launch powers as high as +27 dBm (500 mW), then nonlinearities can come into the picture. Such very large optical transmitter powers are often required because of the roughly 28-dB inferiority in noise immunity of analog TV versus digital TV [Poynton] and the analog multichannel nature of the TV component of triple-play FTTH systems. This is why the 1550 band was allocated to analog TV—this is the only one of the wavelengths at which cost-effective optical power amplification is available (in the form of EDFAs).

In the above list of impairments, the important ones in the FTTH context are loss, chromatic dispersion, Rayleigh and Brillouin scattering, with Raman scattering, cross- and self-phase modulation and four-wave mixing of minor concern. If we were talking about the interoffice or long-haul environment, the relative importance of these impairments would be totally different.

We shall now discuss the FITH-related ones in sequence.

3.2.1 Chromatic Dispersion

Chromatic dispersion is what happens to the propagating signal because of a difference in propagation velocity across the signal bandwidth. It is a property of the glass material and not of the shape and dimensions of the fiber core, so it should not be confused with modal dispersion. Chromatic dispersion is a natural property of all glass, including optical fiber, and one that explains the familiar ability of a glass prism to split out incident white light into its constituent colors. In communication systems, any kind of dispersion causes different parts of a received wideband signal waveform to suffer different time delays, and this can then even get so bad as to cause intersymbol interference. Dispersion is measured in picoseconds of time smear δ per kilometer per nm of bandwidth. A rough rule of thumb for determining the maximum fiber length L before chromatic dispersion reaches harmful levels is [Green]

$$L \leq 100 \, c / \lambda^2 \delta \, \Delta f$$

where transmitted signal bandwidth in hertz is Δf, c is free-space light velocity in meters per second, and wavelength λ is in meters. The upper curve of Figure 3.4 shows δ as a function of wavelength for silica, and it is seen that for standard SMF, δ passes through zero at a wavelength around 1310 nm. This is why this wavelength has been chosen as the one at which PONs operate to transmit all upstream data and voice (and perhaps video). As we shall see in discussing lasers, those in the subscribers' ONUs, where low cost is so important, can be Fabry-Perot (MLM) lasers, which emit light over a broad wavelength band, rather than the much more expensive **distributed feedback (DFB) lasers**, also known as **single longitudinal mode (SLM) lasers**. If MLM lasers are operated at wavelengths other than 1310 nm, the signal received at the head end arrives significantly smeared, in a way that is shown in Figure 3.4*b*.

But we have seen that not only are very high speed signals sent upstream in the form of data near the zero dispersion wavelength of 1310 nm, but very high speed downstream data and video must travel near 1550 nm because that is where the amplification is available. How do we handle the roughly 17 ps/km/nm chromatic dispersion that Figure 3.4*a* tells us occurs at 1.55 μm wavelength?

Several options are open to us [Desurvire-1]. One of the most widely used is to splice a length of **dispersion compensating fiber (DCF)** onto a run of standard

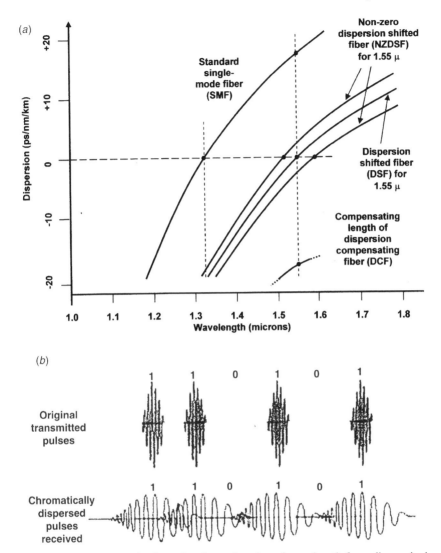

FIGURE 3.4 (*a*) Chromatic dispersion δ as a function of wavelength for ordinary single-mode fiber (SMF), dispersion shifted fiber (DSF), and two choices for nonzero dispersion-shifted fiber (NZDSF). (*b*) Transmitted and received dispersed pulse shapes, respectively.

single-mode fiber (SMF) where the DCF has the opposite sign of chromatic dispersion from that of the SMF. The opposing values of δ available from today's DCF are six to seven times that of the SMF, so the length of DCF needs to be only one-sixth or one-seventh that of the length of SMF. However, there is the attenuation penalty for the big spools of SMF required to do this, since DCF has at least 0.3 to 0.6 dB/km attenuation, compared to 0.20 for SMF.

At least as widespread as DSF is the substitution for classical SMF by **dispersion-shifted fiber (DSF)**. Both DCF and DSF behavior are gotten by shaping the core index profile, not as the conventional simple cylindrical waveguide (Fig. 3.2*b*), but using a more complex radial index profile. This has the effect of moving the zero dispersion point away from the normal 1310 nm to some other wavelength. Several choices for the shift are shown in Figure 3.4*a*. One could pick 1550 nm or create some small offset from 1550 if it is desired to retain a small amount of chromatic dispersion to upset the phase relationships upon which four-wave mixing depends, the so-called **nonzero dispersion-shifted fiber (NZDSF)**.

A third widely used approach is to insert a **chirped fiber Bragg grating (FBG)** component into each fiber somewhere before it reaches the receiver at the head end. An FBG is a segment of fiber or waveguide a few centimeters long, along which has been fabricated a train of index variations whose spatial frequency sweeps across a certain range in a way exactly inverse to the dispersed waveform of Figure 3.4*b*. The transmission fiber and the FBG constitute an example of the familiar **matched filter** pair. Such FBGs can be fabricated to almost exactly compensate chromatic dispersion at one wavelength and one distance. Therefore, at 1550 nm, for example, there must be one FBG part number for each of a number of transmission distances in the system.

3.2.2 Loss and Rayleigh Scattering

Figure 3.5 shows a breakout of the various effects that produce the fiber attenuation curve that was given in Figure 3.1. There are two losses that are intrinsic and unavoidable properties of the glass material. The first of these is the infrared material loss, which is the tail of an absorption loss spectrum that consists of peaks out beyond 9 μm having an attenuation of more than 10^{10} dB/km! Second, there is a corresponding ultraviolet material loss at much shorter wavelengths, and its tail is also shown in Figure 3.5. However, this loss is seen to be greatly exceeded by the Rayleigh scattering loss.

The Rayleigh scattering loss and the bumps in the curve due to vibrational resonances of the OH molecule (water) are the fixable fiber impairments, and the history of optical fiber development has centered around reducing these two. Rayleigh scattering involves light hitting scattering centers whose diameter is smaller than the wavelength λ. It is what makes the sky blue (the λ^{-4} dependency of loss skews toward shorter wavelengths the spectrum of sunlight scattered from microscopic scattering centers in Earth's atmosphere). In fiber, it is caused by tiny density and compositional inhomogeneities. It appears that a limit in reducing these inhomogeneities has now been reached—for about a decade now, the minimum attenuation achieved with production silica fiber has held steady at around 0.15 to 0.2 dB/km.

The OH molecular resonances, on the other hand, have recently been tamed with **ZWP (zero water peak) fiber** to the extent that the bumps shown in Figures 3.1 and 3.4 have been almost completely eliminated. This was done by painstakingly detailed attention to eliminating all possible sources of water in the fabrication process. Today, almost all new SMF bring installed anywhere is ZWP fiber.

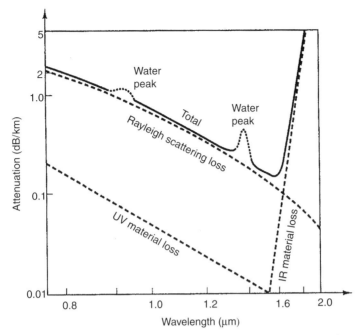

FIGURE 3.5 Details of the physical mechanisms underlying the attenuation curve of Figure 3.1.

3.2.3 Stimulated Brillouin Scattering (SBS)

Stimulated emission will be introduced more completely when we discuss EDFAs. It is what makes lasers lase and erbium amplifiers amplify. Whereas Rayleigh scattering is omnidirectional, SBS [Green, fiber-optics info] is highly directional in that the directivity pattern of the scatter has a null in the forward direction. Much of the scatter hits the side of the core and vanishes into the cladding, but enough propagates back to the laser to constitute a very harmful backreflection, and this can have two effects:

- Destabilization of the laser's operation. As we shall see in Section 3.6, there is a parameter called **relative intensity noise (RIN)**, the part of a laser's output that is noisy rather than monochromatic. RIN depends critically on how much backward-traveling signal enters the laser from downstream. Also, and perhaps more importantly, there is.

- Loss of received signal by conversion of the laser output into stimulated Brillouin scattering over some of the fiber path between laser and receiver.

The effect of SBS on the transmitting laser is exacerbated by the fact that with SBS, the backreflection is always downshifted at a frequency that is very close to the laser frequency (only 11 GHz or 0.09 nm at 1550 nm).

This 11 GHz is intrinsic to SBS. The scattering occurs from quasi-particles called **phonons**, which are the sound wave analogs to photons. A phonon can be viewed as a virtual grating formed in the fiber material and receding at the fiber's sound velocity. Thus, if one takes the ratio of the medium's sound velocity to its light velocity and then multiplies by the frequency of the incident laser light, one gets the 11-GHz downshift. The SBS phonons are formed at high-power levels in the fiber, power sufficient to form the virtual grating in refractive index of the material. Unlike Rayleigh scattering, the fraction of light subjected to Brillouin scattering is not only dependent on incident power but also essentially independent of the wavelength of the incident light.

Stimulated Brillouin scattering can be very harmful to systems with high power and long fiber runs, like PON systems at 1550 using EDFAs [Kelly]. As Figure 3.6 shows, when one exceeds the SBS threshold of signal power into the fiber (typically +6 to +10 dBm for 10-km lengths of single-mode fiber), the down-shifted backscattered SBS has become sufficiently large that it stimulates still more backward light (at the same down-shifted frequency). Further increases in transmitted power cause such backscattered light to be turned into more backward light, less is delivered to the receiver, and more goes into increasing the noisiness of the transmitter (increasing its RIN).

Whereas there is nothing one can do to mitigate Rayleigh scattering (except move to a longer wavelength), there are a few things that can be done to reduce

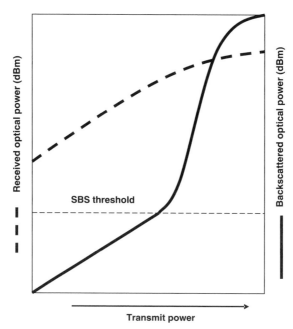

FIGURE 3.6 Showing that above a certain transmitter power threshold, stimulated Brillouin scattering not only steals signal power that should arrive at the receiver (dashed line) but converts it into backward-travelling light (solid line) [fiber-optics info].

the amount of SBS that is incurred [Kelly]. Placing high-powered EDFAs as far downstream in the fiber PON infrastructure as possible is one solution, although unattractive in maintenance and servicing terms. If the laser has been modulated by turning it ON and OFF (**external modulation, direct modulation**, or **on–off keying**), a small AC component can be added either to the amplitude, the phase, or both during the DC "ON" signal during a "1" bit. Power at the modulation sideband frequencies and at the main laser output frequency can all then be reduced below the threshold shown by the dashed line in Figure 3.6. For detectability reasons, it is not desirable to vary the laser output power, so phase modulation has been chosen instead of amplitude modulation.

3.2.4 Stimulated Raman Scattering (SRS)

Like SBS, SRS can also take place with a strength roughly independent of incident wavelength, a fact that has been exploited in developing optical amplifiers using the Raman effect. This effect is very different from Brillouin scattering in that it is wideband and omindirectional rather than backward traveling. The incident light, if it is strong enough, delivers part of its energy to mechanical vibration of the silica molecule and part into reradiated light (Stokes light) of lower frequency than the incident light (because energy has been lost and energy of a photon is proportional to its frequency (energy $E = hf$, where E is in joules, f is optical frequency in hertz, and h is Planck's constant 6.63×10^{-34} J/s). This down-shifted light radiates omnidirectionally, and that which reaches the receiver can introduce significant interchannel crosstalk into the adjacent lower-frequency channels because the spectrum of this SRS component is very broad, as shown in Figure 3.7.

Stimulated Raman scattering is of minor impact in today's FTTH systems for two reasons. First, it is completely dominated by SBS since, for a given incident power level, the Raman effect is three orders of magnitude weaker than the Brillouin

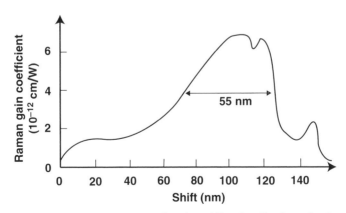

FIGURE 3.7 Spectrum of energy reradiated omidirectionally by stimulated Raman scattering.

effect. The second is the very large wavelength separation between the allocated voice, data, and video channels, which minimizes the amount of Raman interference in adjacent PON channels. As the optical spectrum becomes congested, leading to closer wavelength spacing, interchannel Raman crosstalk could come into play. It is already important as a means of building optical amplifiers because the amplification is available at essentially any wavelength [Islam]. It would certainly be a factor in WDM PONs. The ITU specifications for interchannel separation in such systems are numbers such as 100 GHz (0.8 nm at C-band).

In today's PON systems the SRS phenomenon is most likely to cause problems when the power to launch digital data and voice downstream at 1490 nm becomes large enough to induce SRS 60 nm longer at 1550, the wavelength used for analog TV, which is very sensitive to distortion and noise. For modest, unamplified transmitters at the head end (say up to 2 mW from a laser diode, not an EDFA, as might be the case at 1.25 Gb/s), the degradation in receiver SNR will be less than 0.25 dB [George].

3.2.5   Self- and Cross-Phase Modulation (SPM and CPM)

These effects also occur at high power, when the optical signal in the fiber becomes so strong that the largest positive sinusoidal excursions cause a small increase in index of refraction. This can slightly retard the high-power part of the signal waveform that caused the index change, constituting **self-phase modulation (SPM)**. It can also cause a phase change in some other signal traveling on the same fiber at some other wavelength, **cross-phase modulation (CPM)**. SPM and CPM cause different parts of the waveform to be retarded differently, introducing a frequency chirp and broadening the received pulse. While not yet a recognized problem for PON distances and bit rates, SPM and CPM have troubled fiber transmission systems operating at high power and longer distances. Among the mitigation measures are the use of fiber with a large effective core area, which is also an antidote for the other nonlinearity impairments.

3.2.6   Four-Wave Mixing (FWM)

This is a fancy term for what radio engineers or hi-fi buffs would know as third-harmonic distortion. It is the optical equivalent of the RF impairments CSO and CTB that we discussed in Section 2.5.7. Second-harmonic distortion produces products that lie far from the region of interest, but third-harmonic distortion, when channels are equally spaced in frequency (as with a dense WDM system), can produce cross products at frequencies $f_i - f_j + f_k$, which, when i, j, and k are consecutive integers, is none other than f_j. This effect is not expected to be a problem for any PON not using dense WDM.

■ 3.3   OPTICAL AMPLIFIERS

Another topic relevant to fiber propagation needs to be discussed here, and that is the role and limitations of optical amplification. As mentioned in the last chapter,

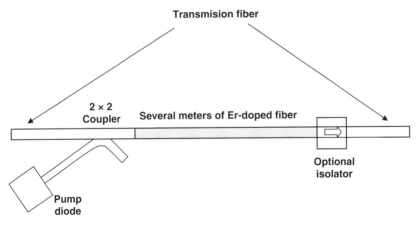

FIGURE 3.8 Schematic of an erbium-doped fiber amplifier (EDFA).

not only is there the inevitable fiber attenuation in the path between head-end OLT and subscriber ONU, but there are also the equally inevitable incidental losses in the connectors and splices. And there are even larger losses incurred in splitting the signal as is done in the PON architecture. For example, the 32 : 1 splits mentioned as examples in the G.983, 4, and 802.ah standards discussed in the last chapter impose 15-dB splitting losses. The **excess loss** that every coupler or splitter has typically ranges from 1.2 dB for a 1 : 4 and 2.0 for a 1 : 8 device. So for the sample case of a PON with 1 : 4 and then 1 : 8 splits that was shown in Figure 2.3, there is typically $9.0 + 6.0 + 1.2 + 2.0 = 18.2$ dB of total coupler or splitter loss on both outgoing or incoming optical paths that need to be made up by some form of optical amplification. And these numbers do not even include the small fiber propagation loss (typically at least 0.2 dB/km) and excess losses in nonideal connectors or splices.

Convenient and economical optical amplification is available today from the erbium-doped fiber amplifier (EDFA) of Figure 3.8. However an EDFA amplifies only in the range from about 1530 to 1560 nm. Interestingly enough, the EDFA is itself made out of fiber, not out of some other kind of optical technology such as lenses or mirrors, and this has many advantages [Desurvire-1]. For example, getting the light into and out of the device is simply a matter of splices or couplers, not more elaborate beam-shaping artifices.

The way such an amplifier amplifies is shown in Figure 3.9. Increased energy of one of the outer erbium atom electrons reads vertically. A **pump laser**, typically at 980- or 1490-nm wavelength, excites the atoms in the erbium impurities that have been deliberately introduced (doped) into a few meters of conventional silica fiber. This pump laser kicks an outside electron of an erbium atom into a higher energy state, as shown at Figure 3.8*a*. A particularly effective way of causing the electron's energy to decay back into the state from which it came is not to wait for it to do so **spontaneously** (Fig. 3.8*b*), but involves **stimulating** it, as in Figure 3.8*c*; tickling it with in incident photon whose energy $E = hf$ exactly matches the energy difference from its present excited state back down to one of its original states.

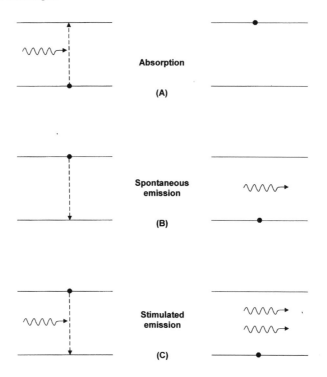

FIGURE 3.9 Amplification of light. Energy diagram showing conversion of photons into energy and vice versa: (*a*) absorption, (*b*) spontaneous emission, and (*c*) stimulated emission.

Thus, the amplification occurs because, for each such occurrence, one gets "two for the price of one"—two output photons for every incident photon (including the incident one). And the gain of an EDFA is not just limited to 3 dB (factor of 2 in energy) because, as with the fission product particles in a nuclear reactor, each of the secondary photons in turn can trigger off another such 3-dB event, and so forth. EDFAs with gains up to 20 dB have proved practical, have been steadily dropping in cost, and have been packaged in volumes as small as 200 cm³. The trouble is that, despite all attempts to broaden out the range of frequencies over which the stimulated emission takes place, the largest bandwidth coverage that can be gotten from EDFAs is about 35 nm, enough to cover C-band (Fig. 3.1). A small part of this broadening of the natural line spectrum of the electron transition occurs by virtue of the fact that the erbium atoms are imbedded in a glass host, whose molecules are at an assortment of distances from the various erbium atoms, and thus exert various amounts of pulling of their natural transition frequency *f.* Even more bandwidth broadening is done by introducing further impurities in addition to the erbium (co-doping it), notably with alumina (aluminum oxide).

To be sure, there are other physical phenomena that provide optical amplification, notably the Raman effect, which, as we saw, requires huge pump powers, on the order of watts, but can be made to amplify at essentially any wavelength.

However, the workhorse optical amplifier today is the EDFA. Fortunately, as we saw in the last chapter, the architects of APONs and EPONs have allocated the best EDFA wavelength, 1550 nm, to that component of triple-play requiring the highest received SNR, namely outbound multichannel *analog* TV. If the TV is digital, then 1490 is used for all three outbound components (remember that *digital* TV requires 28.5 dB less SNR than does analog TV). The EDFA in an FTTH system usually takes the form either of a **power amplifier** at the head end or a **line amplifier** somewhere out in the outside-plant part of the PON, or both.

■ 3.4 SPLITTERS AND COUPLERS

Whereas such older last-mile technologies as DSL, cable, HFC, and FTTC have important powered electronic elements between head end and subscriber, an FTTH system, whether a P2MP PON or a P2P star, has none. The only active element might be an EDFA, and this is not only optically transparent but also protocol transparent. In PON systems, instead of active elements, one uses passive **splitters** (which, considered in the reverse direction, are **couplers** or **combiners**). But these components, while dissipating no electrical power, nonetheless add attenuation and cost issues that have to be considered.

It is easy to underestimate how inexpensive complex electronics have become, thanks to large-scale integration, whereas such techniques have had only limited impact on the cost of optical components, including splitters. The failure of dense spatial integration to be of much help in the optical communication world can be traced to the physics of light propagation in waveguides. A standard fiber or waveguide at, say, 1300 or 1550 nm may not have a bend radius of less than (typically) 2 to 3 cm [Chomycz] or the radiation in the core (particularly at longer wavelengths) gets coupled into modes that propagate in the cladding and then get lost. The effect is more pronounced the longer the wavelength.

Splitters are made today typically either in the form of **fused biconical taper (FBT)** devices or as **planar lightwave components (PLCs)**. FBT devices are made by wrapping several fibers into a bundle, and then, while a flame heats a certain length of the bundle, tension is exerted so that the fused section merges into a narrower bundle of fibers whose cores have been thinned out and brought very close together. PLC components, on the other hand, are typically made by lithographing silica waveguides on metallic silicon substrates in the same style (and often using the same fab facilities) as electronic silicon-based LSI.

In both cases, the principle of operation is the same [Tekippe], as shown in Figure 3.10 for the special case of two inputs and two outputs (a 2 × 2 coupler/splitter). Within each fiber or waveguide there is a long tapered section, then a uniform section of length Z where they are fused together (for the FBT) or lithographed to be narrower and closer together (for the PLC device), followed by a positive taper back to the original cross sections. The tapers are very gradual (**adiabatic**), so that a negligible fraction of the energy incident from either input port is reflected back—that is, the device has a very large **return loss**. As we mentioned in Section 3.1, and as the diagram indicates, narrowing the core destroys the confinement of the radiation

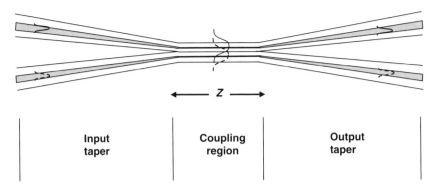

FIGURE 3.10   Principle of operation of a 2×2 coupler.

to lie within it. The field becomes squeezed out into the cladding (including air) and the two fields overlap heavily, as shown.

When the core of a single fiber has been thinned to squeeze energy out of the core, although E and H are in phase in the core, in the cladding they are in phase quadrature. If the evanescent "tails" outside each of two cores overlap heavily, it can be shown [Tekippe] that the two-waveguide structure will support two possible solutions, one having symmetric energy distribution between the two and the other antisymmetric. Fortunately, these two have different propagation velocities. This allows us to control whether or not there will be a strong transfer of energy at a certain wavelength by adjusting Z. The power coupling coefficient is

$$\alpha^2 = F^2 \sin^2\left(\frac{CZ}{F}\right)$$

where F expresses the effect of the difference of core radii in the coupling region, and C the coupling between the two cores within that region. Since α^2 is a raised sinusoid, there are many tricks that can be played so as to pass or reject various incoming wavelengths by adjusting Z. We can see to it that incoming power is split 50 : 50 between the two outputs over a range of wavelengths, thus forming a standard 2×2, or we can maximize the ratio of power at one wavelength to that at another for one output port, while at the same time minimizing it at the other port, thus forming a crude, but inexpensive wavelength division demultiplexor. For example, at the PON receiver it is often necessary to split off the 1550-nm analog video from the digital data at 1490 nm.

It is an interesting peculiarity of single-mode optics that a combiner will have the same loss as the same device used as a splitter. It is not true that the inputs to all branches of the combiner appear with unaltered strength at the output.

As we saw in Chapter 2, for flexible service in passive optical networks, splitter/coupler components are needed with split ratios of from 2 up to 32. For FBTs this is most conveniently done by arranging 1 : 2's in a binary tree, as shown in Figure 3.11, whereupon the entire tree is packaged in a rather large can (typically $1 \times 10 \times 15$ cm for 1×8's). Here is where the alternative PLC technology has a significant advantage. In either realization, one has the option of implementing the 1 : 2 as simply a Y waveguide split.

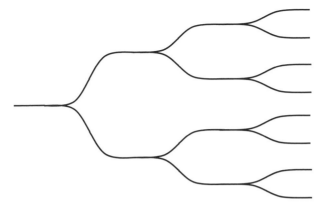

FIGURE 3.11   Making a 1 : 8 splitter as a tree of 1 : 2 fused biconical tapered 2 × 2's.

Typically, FBT devices exist as 1 : 2, 1 : 3, and 1 : 4 units and have a maximum insertion loss of 0.3 dB, a return loss of more than 55 dB, and a rejection in each channel of the other channel's signal of more than 20 dB [FBT]. PLCs are available with splits up to 1 : 32 but require coupling between fiber and PLC waveguide both going in and coming out of the device, giving a slightly higher excess loss than that available to FBT devices using fusion splices. Although at present PLCs and FBT devices with the same specifications are of about equal cost for small port counts, PLC costs are dropping rapidly and are already cheaper at large port counts. For the future, PLCs show the greater promise for significant further cost reductions. Their small physical size has also made it possible to place them in the overhead pods rather in ground-level enclosures in recent Japanese FTTH installations.

■ 3.5 CONNECTORS AND SPLICES

Many forms of single-mode fiber connector have evolved over the years, with ever-improving cost, usability, attenuation, return loss, and suppression of polarization effects. A typical connector type, FC, which is one of those widely used in PONs, is illustrated in Figure 3.12. The fiber is secured with an epoxy adhesive in the ferrule, through which has been bored a long alignment hole of diameter slightly greater than that of the cladding (typically 125 μm). Then the end of the fiber is cleaved, which produces an almost mirrorlike surface, after which the ferrule end, with the fiber inside, is polished.

Since the mating of the two connector halves is never without some residual reflection, often special measures are needed to minimize this echo, whose strength relative to the incident signal is the return loss. For this purpose, many connector types, including FC, are available with the facet not perpendicular to the axis but canted at 8°—the mating connector, also with an 8° facet, must align with this azimuthally. Any residual reflected light will leave this facet at 16°, which is cleverly chosen to be just large enough so that, at the wavelengths of FTTH relevance, internal reflection in the fiber is not supported, and the reflected light is lost into

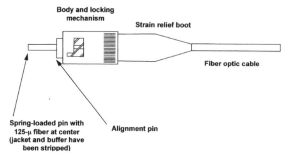

FIGURE 3.12 Geometry of an FC connector.

the cladding. Angling the facet can increase the return loss typically by 15 dB. Angled facets introduce a slight polarization dependence, which exacerbates **polarization mode dispersion (PMD**—a difference in propagation velocity of the two orthogonal polarization states supported by the fiber). While PMD is an issue with long-haul or undersea systems, it has not been of concern for FTTH. Connector types with angled facets are designated with the suffix "APC," (angled physical contact), for example, **FC/APC.**

Fiber-optic connectors have several problems that cause one to use splices wherever there is little likelihood that any rerouting or reconnection will be required. Not only are connectors much more expensive, but the losses are higher, and they cannot easily be ganged to handle fiber ribbons in the style of a multiconductor electrical connector.

Typically, FC connectors can achieve 0.2 dB of loss in the straight-through configuration and slightly higher in the APC form, with 50 and 65 dB of return loss for straight-through and APC cases, respectively. In general, one may expect small connector loss increases with age, particularly from the intrusion of microscopic particles into the interface region.

The urgency of being able to minimize loss in the connector is not just a matter of delivering more power to the receiver. Any connector near the output of a high-powered source, such as an EDFA, is subject to severe damage, which essentially means that the link goes dead.

There are two kinds of splices, **mechanical splices** and **fusion splices** [Desurvire-1]. Mechanical splices [Splices] come in many geometries, but they all share the common features of an alignment structure to ensure exact axial alignment of the two cores, the use of epoxy adhesive to fix the fiber in place, and the possible use of index-matching gel between the fiber ends, plus supporting strain relief and environmental protection. The gel option allows the splice to be disassembled. Fusion splices, on the other hand are permanent. They are made in special hand-held portable units that bring the fiber ends into exact juxtaposition (often with computer-vision software aids using a microscope–camera combination), after which an electric arc is fired, which melts the glass in both fiber ends and fuses them together. Attenuation is usually not actually measured but is instead inferred from a simulation or from measurements on the image. Preceding either a fusion splice or a mechanical splice, it is necessary to **cleave** the fiber exactly perpendicularly to its own axis. Thanks to the existing stress axes in the

FIGURE 3.13   Commercial 12-way fusion splicer for fiber ribbon cables composed of 12 single-mode fibers. The screen shows the 12 fibers. [Ericsson].

fiber, a cleave by a sharp knife edge at a place previously lightly scored, usually produces a flat perpendicular surface. Figure 3.13 shows a ribbon cable fusion splicer.

About the lowest loss that can be achieved by mechanical splices with index matching gel is 0.2 dB for a single fiber and 0.4 to 0.8 dB for a 12-fiber ribbon. With fusion splicers, the numbers are much better: 0.02 and 0.04 dB, respectively.

■ 3.6   LASERS AND TRANSMITTERS

Laser history is interesting. The term has been around for decades and literally refers not to emission but to amplification ("light amplification by stimulated emission and reflection"). This would give one the wrong impression—that light amplifiers evolved before laser emitters. Actually, although the semiconductor laser light *source* was invented in 1965 and is now widely used for everything from welding to eye surgery to CD and DVD players to optical communications, it was 20 years before practical light *amplifiers* were invented in the form of EDFAs.

We have mentioned that EDFAs work to amplify the incoming signal by causing it to impinge on erbium atoms that have been *pumped*, meaning that their outermost electrons have been moved to a higher-energy orbit around the nucleus. They sit in that energy state until an incident signal of just the right energy (i.e., frequency) *stimulates* them to drop into a lower energy (more interior) orbit, as was shown in Figures 3.9*a* and 3.9*c*. This stimulated emission phenomenon can be observed in many other media besides glass that has been doped with impurity atoms as in an EDFA. In particular, a wide variety of usable semiconductor materials are available. And, by placing two facing mirrors around the region of the material in which stimulated emission is occurring, one can create a source of radiation, a laser, by having any stray optical waves at any of the many resonant frequencies of the cavity formed by the two mirrors become repeatedly reinforced by a cascade of many successive stimulated emission events.

But how is the stimulated emission produced? The answer is shown in Figure 3.14 (the *P–I* curve). With zero pump signal, the medium is essentially opaque, but as the pump intensity increases, a state of **transparency** is reached at which there are frequent enough random **spontaneous** transitions of the electrons from higher to lower energy levels (Fig. 3.9*b*) to exactly compensate for the losses of light in the material. The photons that are emitted are not organized at all but have different frequencies, propagation directions, and polarizations. Then as the pump energy is further increased, the number of these spontaneous events increases to the point that some useful light comes out of the device, which is now acting as a **light-emitting diode (LED)**. There is still nothing to organize these emissions, and so they too fly off in different directions with different wavelengths, and polarizations—they are said to be **incoherent**. These spontaneous emissions are occurring at any wavelength at which the material is exhibiting gain. As Figure 3.14 indicates, the LED spectrum is essentially the **gain curve**. The gain curve in a laser made of the semiconductor indium phosphide, which is the material of choice for FTTH systems at either 1310 or 1510 nm, is very broad, 30 to 40 nm. This spectrum is too broad to be useful for communication systems when chromatic dispersion is present in the fiber, and crosstalk into adjacent channels is to be avoided.

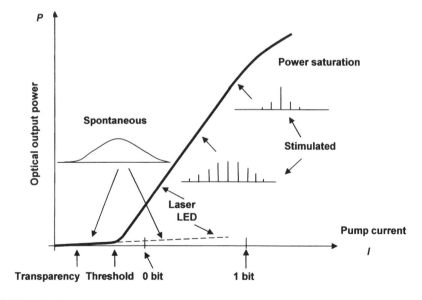

FIGURE 3.14 *P–I* curve of a semiconductor laser.

As the pump level is raised past the point of transparency, the presence of the two mirrors takes effect, and above the so-called **threshold** pump level, the many resonant-mode frequencies of the cavity that lie within the gain curve make themselves apparent. The optical output signal strength begins to be reinforced enough that the former incoherent outputs at the various cavity modes become organized into a set of **coherent** outputs, in other words, each output at a cavity-mode frequency is composed of photons with identical phase, direction and polarization. Because the radiations at each frequency are now in phase, further increases in the pump current lead to very large increases in output, in a proportion that is linear. As the pump increases further, a concentration of the lasing into fewer and fewer radiated modes becomes more pronounced.

Figure 3.14 also depicts typical choices of the drive current for on–off modulation of the laser in sending digital data. Notice that for a 0 bit, the laser is not turned completely off but is **biased** slightly above threshold. The reason for this is that if it had been turned off completely by using a current value for a 0 bit that is less than the threshold current, then it would have taken a long and randomized time to turn it on again, typically up to 10 ns. During this interval, what starts out as spontaneous emission only gradually and unpredictably turns into the desired stimulated emission. Clearly, to be able to send gigabit rates, such **turn-on delays** will not do. As we saw in the specifications for APONs and EPONs in the last chapter, noninfinite **extinction ratios** must be accepted in order to send high-speed data. The extinction ratio is the ratio of 1 bit current to 0 bit current, and is typically 6 to 10 dB. One might also have noticed from the discussion of PONs in the last chapter that when an ONU is not talking or has not started its permitted time slot to do so, the bias *is* in fact reduced to zero. This is to prevent small amounts of spontaneous emission from each of the temporarily inactive ONUs to contaminate the permitted ONU's signal with spontaneous emission noise.

Lasers in which the only resonant structure is the cavity formed by the two facets are called **multimode** lasers [referred to by the ITU as **multilongitudinal mode (MLM) lasers**]. They still will not do for the downstream FTTH transmission at 1490 or 1550 nm. This is because, even at high pump current, there is still a great deal of radiation output at many frequencies, which will interact with the fiber's chromatic dispersion to produce excessive time smearing. On the other hand, such MLM lasers work perfectly satisfactorily at the 1310-nm zero dispersion wavelength of upstream data, thus allowing the laser in the subscriber's ONU to be of low cost. MLM lasers do not require thermoelectric coolers, and this also saves on cost.

What is wanted for 1490- or 1550-nm service at the head end is to suppress all the cavity modes but one. By the way, one should not confuse the two uses of "multimode." In the case of fiber the word refers to multiple permitted standing-wave patterns laterally across the fiber, whereas with a laser it refers to the permitted resonant modes longitudinally along the cavity between the two mirrors.

The method of choice for getting rid of all the cavity resonances but one is to lithograph along the bottom of the laser cavity itself some striations that act as a spatial filter that suppresses all but one resonance spectral line, an artifice called *distributed feedback*. By putting all the pieces of the preceding paragraphs together, we finally arrive at the **DFB (distributed feedback) laser**, whose geometry is shown in Figure 3.15.

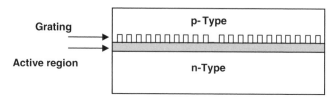

FIGURE 3.15 Geometry of a distributed feedback (DFB) laser.

A small three-layer crystal of indium phosphide is cleaved, which produces mirrorlike flat surfaces at either end. A grating, etched on one side of the middle region provides the distributed feedback. The resulting object is roughly 250 μm on a side, about the size of a grain of salt. Electrodes are attached at top and bottom to inject the current drive into the sandwich of semiconductor alloy we shall discuss in more detail in Section 3.8. The pump current is the abscissa in Figure 3.14. Adjacent to one facet is placed the end of a fiber pigtail to carry off the light at the single output frequency, and the lens that focuses the light from the facet into the pigtail is simply the curved end of the fiber, formed by melting the fiber end in a certain way.

At the other facet there is usually placed a **monitor photodiode** whose function is simply to provide a diagnosis of whether the laser is radiating the power it should. It is also necessary to include a **thermoelectric cooler (TE)** to maintain the laser chip's temperature low and constant in order that the threshold current value be kept low and constant. A TE cooler uses the Peltier effect by which current is converted to energy removal, that is, temperature drop. An external closed-loop control circuit uses the voltage from an internal **thermistor** to sense the temperature and drives the internal TE cooler to control it. Packaging the entire structure of laser, coupled pigtail, monitor photodiode, thermistor, TE cooler, and electrical connections into a small can about 1 cm on a side completes the picture.

■ 3.7 PHOTODIODES AND RECEIVERS

A **photodiode** form of photodetector is in essence the inverse of a laser, in ways that will be detailed in the next section. Light enters and is converted into current, whereas in a laser a pump current is converted into light. In spite of this seemingly simple inverse relationship, **photodetectors** do not look at all like lasers, for many reasons. For one thing, the current produced is so tiny that it is customary to package a **field-effect transistor (FET)** amplifier in the same component as the photodiode, thus forming a so-called **PINFET** photodetector (PIN for the physics of the photodiode: "positive-intrinsic-negative," as described in the next section). For another, the PIN detector itself must have a much larger physical area in order to collect the light, so that problems of coupling light between fiber and device are orders of magnitude more forgiving. Also, with the photodiode no bias current supply is needed, nor is elaborate temperature control.

Not only is it customary to package the photodiode with an FET preamplifier to get the most SNR out of the photodiode, but if the bit rate is prespecified, it is feasible to add the required threshold decision circuit and even clock recovery to form an entire **photoreceiver**.

We saw in Section 3.3 that the most common method of amplifying a signal at the optical level, rather than the postdetection electronic level, was the use of EDFAs. Note that it is very important to do whatever amplification must be done *predetection*, while the signal is still a photon flow rather than an electron flow. This is because one photon worth of energy converts into one electron worth of current, so that a given number of decibels of optical gain is equivalent to twice that number of decibels gotten later on after photodetection (amperes into watts). This is one of the reasons for the importance of a second form of optical detector, besides the photodiode, namely the **avalanche photodiode (APD).** By causing each incident photon not only to trigger off one electron, as with a standard photodiode, but then to arrange matters so that this electron in turn triggers off one or more secondary or tertiary electrons, one can get up to 10-dB of additional detectability. In the next chapter we shall see this 10-dB difference reflected in Figure 4.2, giving curves of required received signal power versus bit rate to achieve various bit error rates.

■ 3.8 THE PHYSICS OF LASING AND PHOTODETECTION

We conclude this chapter by laying out the physical processes that lead lasers and photodetectors to do what they do. In the earliest lasers, the active medium was a gas, and the pump was a high-intensity light of a certain wavelength. By

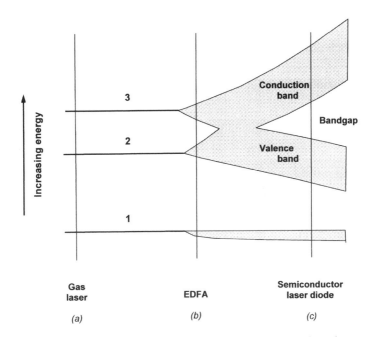

FIGURE 3.16   Showing how narrow energy levels become bands when the active atoms are (*a*) far apart as in a gas laser, then (*b*) crowded next to solid co-doping impurity atoms (as with an EDFA), and (*c*) finally part of molecules in a crystal (as in a semiconductor laser or photodiode).

considering the early gas laser along with EDFAs and semiconductor lasers, we can create a sort of unified picture of the physics involved, starting with Figure 3.16.

- In the **gas laser**, the atoms are very far apart, and the fact that they do not interact results in the laser output occupying a very narrow spectral line. The energy of the outer electron of a gas atom is permitted to occupy only certain levels, of which three are shown in Figure 3.16a. Normally most of the electrons have energy at level 1. Applying a pump signal at exactly the frequency corresponding to the 1 to 3 energy difference raises the energy level of the outer electrons of many gas atoms from level 1 to the highest level 3. As an intrinsic property of the chosen gas, level 3 has a long lifetime before spontaneous decay will occur from 3 to 2. Therefore, many electrons are stored at 3, thus constituting a stable **population inversion** that keeps getting refreshed by leaving the pump on. Now, if a signal photon of the same wavelength (energy) as the 3 to 2 transition comes along, it will simulate the emission of a second such photon (with the same frequency, direction, phase, and polarization), both will in turn stimulate the emission of further such photons, and a powerful optical output beam will be produced.

- In the **erbium-doped fiber amplifier**, Figure 3.16b, the pump energy is likewise in the form of a radiation at a shorter wavelength (e.g., 980 nm) than that of the signal to be amplified (around 1550 nm), and this propels an outer electron into a higher energy orbit, where it awaits the arrival of one of the photons to be amplified. But, unlike the gas laser case, each erbium atom is surrounded by other atoms, including those of the alumina impurity, which, although they may not take part in either the spontaneous or the stimulated emission processes (the latter constituting the amplification), nevertheless act to pull the frequency of the emission by various amounts due to being at varying distances from the erbium atoms because the material is amorphous not crystalline. This results in frequency broadening of the amplification, as shown in Figure 3.16b by the fact that the energy levels are no longer the narrow lines of the gas laser case.

- In a **semiconductor laser diode**, in which the material is a crystal with specific interatomic separations, the process is different and is the subject of the remainder of this section.

As we see from Figure 3.16c, there are in a semiconductor three permitted energy bands that can be occupied by an electron, including the upper **conduction band** (being so-called because electrons having those energies are not associated with parent atoms, and so they can participate in electrical conduction), and the **valence** band (because it is outer electrons of individual atoms, whose energies lie in this band that participate in chemical reactions).

A different view of these two bands is shown in Figure 3.17. Whereas in the gas diode and the EDFA, light does the pumping, in the laser diode pump current comes in, causes electrons in the conduction band to drop in energy to the valence band, and light comes out, stimulating further transitions in the process. As we shall see, in a **semiconductor photodiode**, the process is the reverse—arriving light

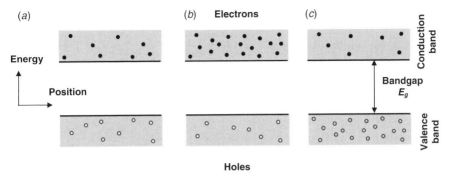

FIGURE 3.17 Valence and conduction bands and bandgap in the absence of applied current, with energy reading vertically and one physical dimension reading horizontally: (*a*) intrinsic semiconductor material, (*b*) n-type, and (*c*) p-type. Black dots and circles are free electrons and holes, respectively.

kicks valence band electrons into higher-energy conduction band states, producing an electron flow in the attached electric circuit.

The conduction band has the property that if there are any unbound **electrons** possessing this range of energies, they are effectively free to move around physically. Reciprocally, the valence band has the property that if there are any unbound **holes** (places where physical electrons are missing) in this range of energies, they are likewise free to move around physically—horizontally in the diagram. Neither is permitted to occupy the region between the bands, the **bandgap**, whose energy span we denote by E_g.

All this is illustrated in Figure 3.17, which shows energy versus one-dimensional position along the device for three types of doping (addition of specific impurities) of the same semiconductor material, **intrinsic** (no doping), **n-type** (**donors** donate electrons) and **p-type** (**acceptors** contribute holes).

Note carefully that this figure and the next two we shall be discussing do not show device shape—while the horizontal axis is distance along the device (along the axis in which current flows, as it will turn out), the vertical axis is *energy*.

If we could just somehow put the electrons at the top at the same horizontal position in this diagram as the holes at the bottom, this would mean that they are in close physical proximity to one another, and they can then combine, releasing a photon of energy E_g and, therefore, of frequency $f = E_g/h$. We now show how they can be made to crowd together physically so that this occurs.

In an intrinsic semiconductor, say pure InP, there are very few conduction band electrons (black dots) or valence band holes (white dots), and the material is electrically neutral. However, if the InP has been doped with an occasional impurity atom of gallium, it then has an excess of free electrons (which are free to move around only in the conduction band) and therefore becomes n-type, or if doped with phosphorus it becomes p-type with an excess of holes (which must lie in the valence band).

Now, consider Figure 3.18, which depicts the simplest kind of semiconductor laser diode, the **homojunction** device. The device shown is a two-layer sandwich

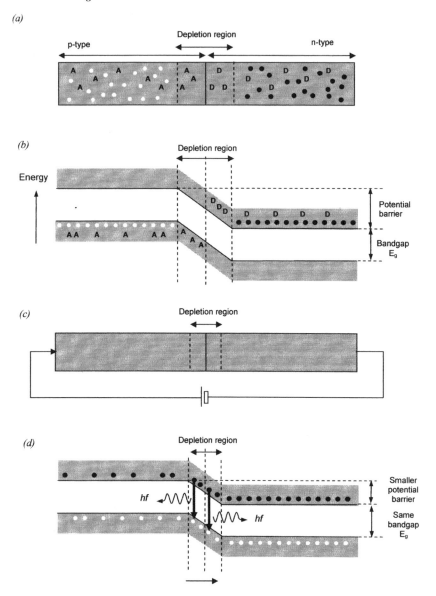

FIGURE 3.18 Energy bands in simple homojunction laser: (*a*) open-circuit device, (*b*) energy vs. position for an open-circuited condition, (*c*) forward biasing, and (*d*) energy vs. position for forward biasing. Small D's represent donor atoms and small A's acceptor atoms.

(reading horizontally) of n-doped and p-doped materials, for example, InP whose two layers have been doped with gallium and with phosphorus, respectively. Current is applied as shown at Figure 3.18*c*. By convention, the electrons flow in the direction opposite to the arrow.

When the two materials are first brought into contact, as shown physically at Figure 3.18*a* and with respect to energy at Figure 3.18*b*, free electrons (ones not associated with parent atoms) from the n-side try to flow leftward to the p-side to even out the density of electrons laterally, and the converse happens with free holes traveling rightward from the p-side to the n-side. However, as these free electrons and free holes pass across the junction, many electron-hole pairs combine instantly, producing radiation and leaving neutrally charged atoms. This leaves the n-region populated with few free electrons but now with a number of donor atoms that have been supplied with extra outer orbit electrons by the free electrons of the drive current (each denoted by D in the diagram), and leaves the p-region with few free holes but extra acceptor atoms (each indicated by an A, each representing a hole). So, all this leaves a region in the middle, the **depletion region**, containing mostly donor atoms and acceptor atoms, not free electrons and holes.

Any further leftward electron movement from the n-region is opposed by the coulomb repulsion (of like charges) between the donor atoms on the right side of the junction and the free electrons further to the right on the n-side. Similarly, the acceptor atoms immediately to the left side of the junction repel the holes on the p-side from moving to the right. In this way an energy barrier is produced, as shown at Figure 3.18*b*, that the electrons are unable to climb over and the holes (for which increased energy is downward) are unable to crawl under. The height of this **potential barrier** is the device's open-circuit **contact potential**.

Note that there is almost no place along the length of the device where there are simultaneously many electrons in the conduction band of energy and at the same place many holes in the valence band. However, if one now applies a forward current, as at Figure 3.18*c*, wherein the electrons flow from the right, the large potential barrier that we had for the open-circuited condition Figure 3.18*b* is reduced, which reduces the coulomb repulsion that was holding the electrons on one side and holes on the other, and the depletion layer is no longer depleted—there are electrons in the conduction band and holes in the valence band *at the same place* along the device, namely near the interface between the n-side and the p-side. The result is shown at Figure 3.18*d*, where holes and electrons *at the same place* combine to produce radiation of the frequency E_g/h.

As we saw with the $P-I$ curve of Figure 3.14, as the current is increased, the potential barrier is reduced still further, more photons are produced, and if there are mirror facets at the ends of the active region (vertically in the diagram), stimulated emission rapidly predominates over spontaneous emission, and lasing starts taking place. If there are no mirrors, spontaneous emission continues to grow in intensity with injected current, thus forming not a laser but an LED.

The converse processes for a so-called **PN photodiode** are shown in Figure 3.19. Going back to Figures 3.18*a* and 3.18*b*, we see the open-circuit condition, as before. Now, as Figure 3.19 shows, sending current through the device in the opposite direction, as shown at Figure 3.19*a*, the depletion region of Figure 3.19*b*, instead of getting narrower becomes even wider than in the unbiased case of Figures 3.18*a* and 3.18*b*. The potential barrier becomes even taller, that is, the voltage difference across the depletion region becomes even stronger, and every time an electron-hole pair is created by an incident photon of $f = E_g/h$ arriving

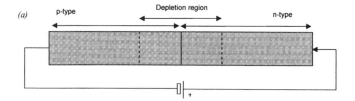

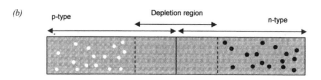

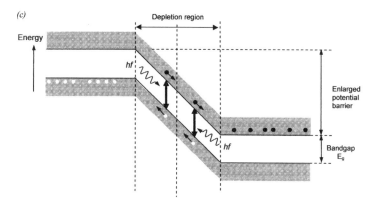

FIGURE 3.19 Physics of a simple PN homojunction photodiode: (*a*) reverse biasing and (*b*) energy vs. position for reverse biasing.

vertically in the diagram, the electron will move rapidly to the right and the hole to the left. In this way a stream of photons is converted into a current.

In a photodiode, we would like the depletion region to be as thick as possible, so that as many of the incident photons as possible will be converted to current flow, rather than passing through the layer unproductively. It does not do much good for the photons to be converted to electron-hole pairs in the two other regions outside the depletion region because there is no voltage gradient there to propel them into the external circuitry; they will stay there until they combine via nonproductive processes or diffuse slowly into the depletion region.

This physical widening of the depletion region can be increased by placing a layer of intrinsic material between the p-region and the n-region, forming a **PIN photodiode**. Most photodiodes used in FTTH systems are PIN diodes, usually with an immediately following electronic FET amplifier, thus forming the PINFET device mentioned earlier.

It is important to remember that the conversion of light into current and vice versa obeys a quadratic law since one electron is equivalent to one photon—*P* milliwatts of incident optical power convert into *I* milliamps of current by the formula

$$I = 1.28\eta P$$

where η is the **quantum efficiency** of the device, the fraction of the photons converted to electrons, typically around 70% for a good InGaAsP PIN photodiode such as would be used at FTTH wavelengths.

We have omitted discussion of **heterojunctions**, which are the basis of most of today's practical laser diodes. Analogously to the PIN photodiode versus the PN photodiode, the **heterojunction laser diode** differs from a homojunction laser diode in having a layer of intrinsic material interposed between n- and p-regions. With the right intrinsic material choice, this intervening layer will have a smaller bandgap than the two other regions, and this considerably lowers the lasing threshold.

■ 3.9 SUMMARY

In this chapter, we have laid some of the physics groundwork for the elements that make up the system and have tried to do so with a minimum of mathematics and a maximum of heuristic reasoning. A more thorough treatment of the details may be found in [Kaminow] and [Green].

REFERENCES

[Chomycz] B. Chomycz, *Fiber Optic Installers Field Manual*, McGraw-Hill, New York, 2000.

[Desurvire-1] E. Desurvire, *Broadband Access, Optical Components and Networks, and Cryptography*, Wiley-Interscience, Hoboken, NJ, 2004, pp. 228 ff.

[Ericsson] Model RSU 12-fiber ribbon splicer. www.ericsson.com/networktechnolgies/literature/Fusion_Splicing_literature

[FBT] Fused Single Mode Fiber Wavelength Division Multiplexer Series, www.etsc-tech.com.

[fiber-optics info] www.fiber-optics.info/articles/nonlinearities.htm.

[George] J. George, Designing Passive Optical Networks for Cost Effective Triple Play Support, Paper T-05, 2004 Fiber to the Home Conference, New Orleans, Oct. 4–6, 2005.

[Green] P. E. Green, Jr., *Fiber Optic Networks*, Prentice Hall, Englewood Cliffs, NJ, 1993.

[Islam] M. N. Islam, Raman Amplifiers for Telecommunications, *IEEE Jour. Sel. Topics Quantum Electro.*, Vol. 8, No. 3, May/June, 2002.

[ITU G.652C] Standard for ZWP fiber.

[Kaminow] I. P. Kaminow and T. Li, *Optical Fiber Telecommunications*, Vols. IV-A and B, Academic Press, San Diego, April, 2002.

[Kelly] M. Kelly, Controlling SBS in Measurements of Long Optical Fiber Paths, Application Note, Agilent Technologies,

[Nonlinearities] Fiber Nonlinearities, www.fiber-optics.info/articles/nonlinearities.htm.

Base Technologies

[Poynton] C. Poynton, Digital Video and HDTV, Morgan Kaufman, San Francisco, 2003.

[Splices] www.tpub.com/neets/tm/108.6.htm.

[Tekippe] V. J. Tekippe, Passive Fiber Optic Components Made by the Fused Biconical Taper Process, in *Fiber and Integrated Optics*, Taylor and Francis, UK, Vol. 9, 1990, pp. 97–123.

a VOCABULARY QUIZ

Discuss not only what these terms abbreviate but also what they mean.

Acceptor atoms
Adiabatic coupling
APD
Bandgap
Biasing
Brillouin scattering
C-band
Chromatic dispersion
Cleaving
Coherent radiation
Combiner
Conduction band
Connector
Contact potential
Core
Coupler
CPM
Cross-phase modulation
DCF
Depletion region
DFB laser
Direct detection
Direct modulation
Donor atoms
DSF
EDFA
Electron
Evanescent wave

Excess loss
External modulation
Extinction ratio
FBG
FBT coupler
FC/APC
FET
Fiber nonlinearity
Four-wave mixing
Fused biconical taper device
Fusion splice
Gain curve
Gas laser
Heterojunction
Heterojunction laser diode
Hole
Holey fiber
Hole-assisted fiber
Homojunction
Homostructure
Incoherent radiation
Intrinsic material
Lasing threshold
LED
Loss
Matched filter
Mechanical splice

MLM laser
MMF
Modal dispersion
Mode
Mode field diameter
Multimode laser
n-type material
NZDSF
O-band
On–off keying
Phonon
Photodiode
Photodetector
Photon
Photonic bandgap fiber
Photoreceiver
P–I curve
PIN photodiode
PINFET
PLC
PMD
PN photodiode
Population inversion
Potential barrier
Power amplifier vs. line amplifier
p-type material
Pump laser

Quantum efficiency
Raman scattering
Rayleigh scattering
Return loss
RIN
S-band
SBS
Self-phase modulation
Semiconductor laser
Semiconductor photodiode
SLM
SMF
Splitter
SPM
Spontaneous emission
SRS
Stimulated emission
Thermoelectric (TE) cooler
Threshold
Transparency
Turn-on delay
Valence band
ZWP fiber

CHAPTER 4

Deploying the System

■ 4.1 INTRODUCTION

So far, we have been discussing the overall view of a fiber-to-the-home system as an abstract block diagram or in terms of individual components. It is now time to see how the components can be assembled in actual installations to form the typical FTTH system, such as one would actually encounter out in the real world, tracing it from the OLT at the central office or head end all the way to the user's plug in the wall.

Figure 4.1 shows the nomenclature for the various parts of a PON (P2MP) or a star (P2P) as arranged physically. At the central office (telco lingo) or head end (cable lingo) there is an OLT for each fiber. Each **feeder** cable (containing a number of fibers) is routed to one **fiber distribution hub (FDH)** or **local convergence point (LCP)**, where a split into several **distribution** cables is made in order to connect to subsidiary FDHs or to the smaller housings in aerial splice enclosures called **network access points (NAPs)** or **terminals**. Beyond the terminals are the drops leading directly to the ONUs at individual subscribers' premises. Onto this real-life diagram of an FTTH network can be mapped the PON or P2P block diagrams of Figure 2.1 with the splitting points placed in the FDHs, various convergence points, and NAPs.

The feeders, distribution cables, and drops may be implemented in a rich variety of ways, but basically there are two classes, aerial and underground, which we shall break out into individual methods later in this chapter. Similarly, the convergence points and NAPs can be realized in a variety of **cabinets**, either at ground level or in vaults, or in **closures** (pods) overhead, and we shall discuss these too. The cabinets contain fiber terminations, patch panels, splitters, test access points, and other resources.

It is customary to house the convergence points in cabinets, whereas it is the NAPs that one sees either as closures hung from strength members running between utility poles or in small cabinets at ground level.

Fiber to the Home: The New Empowerment, by Paul E. Green, Jr.
Copyright © 2006 John Wiley & Sons, Inc.

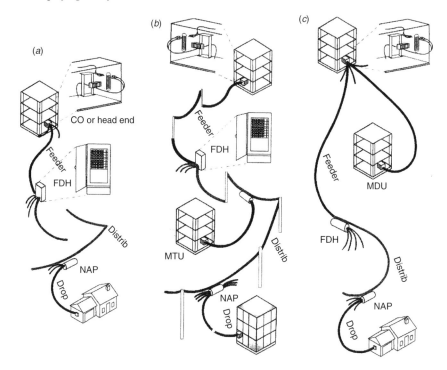

FIGURE 4.1 Maps of typical FTTH systems: (*a*) to an individual residence, (*b*) to a residence and an apartment building, and (*c*) to small businesses. CO = central office or head end, FDH = fiber distribution hub, NAP = network access point, MTU = multitenant unit, and MDU = multiple dwelling unit [Corning].

■ 4.2 THE LINK BUDGET

In Chapter 3 we were at some pains to discuss the various losses incurred in propagating the signal along a length of single-mode fiber cable and other losses that are incurred in splitting the signal (as in a PON), splicing a fiber or cable, or passing it across back-to-back connector halves. We also discussed how different components of the triple-play mix travel at different wavelengths, which in turn have different propagation parameters—attenuation, dispersion, and the effects of nonlinearities. Now it is time to put these numbers together for some real situations.

The **link budget** is a calculation in which some of the actual systems parameters are input and from these others are determined. For example, one can input all the signal power levels, losses, and gains, noise power levels, and determine as the output the bit error rate that would be predicted. As another example, one could specify everything but the distance and then determine the longest length of a PON that could achieve or exceed the desired performance target with the specified optical and electronic component resources at hand.

To do such calculations completely, it is necessary, in principle, to consider:

- The important signal-related items
  - The **bit rate**
  - The target **bit error rate**
  - The **transmitted power** at the designated wavelength for a "1" bit
  - **Extinction ratio**, the ratio of laser power for the ON condition ("1" bit) to that of the OFF condition ("0" bit)
  - Fiber **attenuation** at that wavelength (dB/km)
  - **Splitting losses** in couplers and splitters
  - **Excess losses** in connectors, couplers, and splices, usually given in the spec for the device
  - Amplifier **gain**
  - The **quantum efficiency** η of the photodetector, the fraction of incoming photons that get converted to electron–hole pairs
- The important noise-related items
  - **Thermal noise** in the resistive part of the input impedance of the electrical amplifier that follows the photodiode
  - **Amplified spontaneous emission (ASE)** generated inside any optical amplifier that appears on the path between laser and photodiode
  - **Crosstalk** from signals at other wavelengths, often due to fiber nonlinearities
  - In the case of nonmonochromatic laser sources driving single-mode fiber, as in the return path at 1310 nm in a PON, there is **mode partition noise** as the radiation hops around from one of the permitted cavity resonance modes to another
- Plus a few decibels of **link margin** or margin of safety, left as a hedge against aging effects, temperature changes, or other environmental disturbances, including such things as particles entering the connectors

It does not always have to be this complicated to get reasonable answers quickly. We can start to pull all this together by taking the first two of this list of parameters and using Figure 4.2 to determine how much optical power needs to reach the receiver. (This figure was arrived at by considering typical commercially available photodiodes and avalanche photodiodes, and assuming reasonable values for extinction ratio, photodiode quantum efficiency and thermal noise in the receiver amplifier, while ignoring momentarily any ASE, crosstalk, and mode partition noise.)

From the required "1" bit received signal level required, we can work backwards toward the transmitter, adding 14.5 dB, if there is a total split of 32 in the PON (either in one 1 : 32 split or cascaded 1 : 8 and 1 : 4 splits or any other combination), adding all the decibels of excess loss for the splitters, splices, connectors, and patch panels, and finally arrive at the number of decibels relative to 1 mW (dBm) that the transmitter would have to radiate for a 1 bit if it fed directly into the receiver with only the fiber splitters, splices, connectors, and patch panels in between. Subtracting this from the actual decibels relative to 1 mW of power

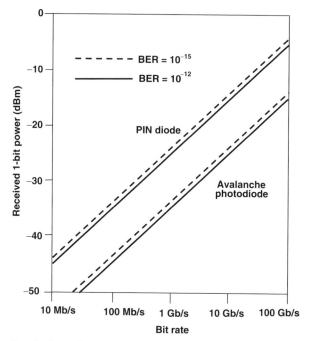

FIGURE 4.2 Required receive minimum optical power during a "1" bit for a typical PINFET receiver, as a function of bit rate for two bit error rates.

available from the laser, and plugging in the fiber attenuation at that wavelength, will tell us how long the fiber run can be between head end and premises before there is insufficient received power to achieve the target bit error rate.

It is useful to see what the standards bodies have to say about link budgets. The numbers have been given in Section 2.5.1 for the BPON, Section 2.6.1 for the GPON, and 2.7.1 for the EPON. As an example, let us plug in some numbers from the Chapter 2 discussion of the EPON standard and the Chapter 3 discussion of the basic technology building blocks. The example is an EPON (Section 2.7.1) with a single 1 : 32 split, 1.0 mW of "1" bit laser power and a bit rate of 1.25 Gb/s at 1490-μm wavelength in the outbound direction. The standard, and also Figure 4.2, show that at 1.25 Gb/s, a typical commercial PINFET requires -24 dBm of received power level during each "1" bit, in order to achieve a 10^{-12} bit error rate.

15.05 dB of that 24 dBm is accounted for by an idealized 32-way splitting loss

0.3 excess loss occurs in the splitter, another $4 \times 0.07 = 0.28$ dB from four fusion splices, and another $3 \times 0.75 + 3 \times 0.5 = 3.75$ dB loss in six connectors, including those at the FDH and NAP (Fig. 4.1), plus 3 dB of link margin, giving a total of

22.38 dB for fixed losses, not including fiber attenuation

In this example, with 0 dBm of transmitter power, we have used up almost all our margin for things other than propagation loss, there being only $24 - 22.38 = 1.72$ dB left for propagation. However, because of fiber's low attenuation of, say 0.172 dB/km, this gives us a distance of 10 km, the number given in Section 2.7.1.

One lesson from this exercise is that the N-way split among the subscribers dominates everything, including the propagation loss, which is the usual worry with conventional fiber-optic systems. At 1550-nm wavelength, for which the ITU G.652 standard for the older single-mode fiber specifies 0.25 dB/km, the PON would have to span 60 km in order for attenuation to have produced as large an effect as a 32-way split does. For newer fiber, the distance would be even greater. As we saw in Chapter 2, the maximum distance for PONs is defined by the time interleaving process, not by distance attenuation. The same is true for DOCSIS cable systems.

Suppose, in our example, that we wish to reach subscribers out to a distance of 20 km using older fiber with 0.25 dB/km attenuation. The fiber attenuation will then be $20 \times 0.25 = 5.0$ dB, so that $5.0 - 0.172 = 4.82$ dB is to be added to the power budget figures given in the example, so in that case, 4.82 dB of improvement must be provided, for example, by increasing the 1.0-mW laser power level into the fiber to at least 4.82 dBm (3.04 mW), a power level unlikely to cause any of the nonlinear effects discussed in Chapter 3 [George].

Downstream analog video at 1550 nm is another matter: 1550 is not a zero dispersion wavelength as is 1310, which has been reserved for upstream data to drive down the cost of the consumer-level laser required at the OLT by allowing it to be spectrally broad. But 1550 is the wavelength where erbium amplification is available. We have seen that amplification is needed because analog systems, particularly multichannel TV, require so much greater received SNR than do the corresponding digital systems.

Whereas in binary digital systems only 15.5 dB of SNR (factor of 36 in power) are required for bit error rates below 10^{-9}, in an analog TV system the SNR must be above 44 dB in order to meet the FCC's minimum requirements and over 47 dB to reduce observable "snow" to invisible levels [Poynton]. Differences such as $44 - 15.5 = 28.5$ dB can only be made up by launching more power or strongly amplifying it en route. In the absence of special SBS suppression measures discussed in Section 3.2.3, up to +13 to +17.5 dBm of launch power (20 to 50 mW) can be used before SBS levels become harmful, if the total path length in the PON is no greater than 20 km [George].

■ 4.3 AERIAL DEPLOYMENT

The most visible and ubiquitous way of deploying nonwireless communication resources has always been by cables stretched between utility poles. Typically, once the power company has erected the poles and taken ownership of the top few feet of vertical real estate on these poles, telcos, cable companies, broadband providers, and others have found the flexibility and low initial cost of aerial deployment irresistible. The rest of the citizenry has found aerial facilities resistible.

In the United States, approximately 32% of all copper access lines are aerial, 53% buried underground outside of ducts, and 14% underground in ducts. Of the

176 million U.S. access lines 75% lead directly to the central offices, and the other 25% are fed by digital loop carriers that connect to the CO by T-carrier lines (1.5 Mb/s) [Jefferies].

Figure 4.3 shows a scene in a part of Tokyo where a number of services, some by fiber, occupy the same utility pole right-of-way, while there is plenty of bandwidth in one fiber to accomplish all but the power. The figure makes the case rather vividly that combining many communication services, each of a modest bandwidth onto one fiber path has not only economic advantages but aesthetic advantages too. There are certain to be monopoly control considerations too, which should keep telecomm lawyers happily occupied as copper gets replaced by fiber on a broader scale.

The low first cost and ease of making upgrades and changes have historically prevailed over the unsightliness of aerial facilities, the environmental extremes to which they are exposed (-25 to $+65°$C, compared to 0 to $+30°$C for underground [Kaminow]), and their greater vulnerability to the mechanics of weather and traffic accidents. A recent comparison of aerial vs. underground costs for fiber subscriber loops shows a 2 : 1 first-cost advantage in favor of the aerial alternative [Render]. Although this huge cost advantage is being steadily eroded by technology improvements in underground installation methods (as we shall see in the next section), for FTTH providers the preexistence of aerial pathways, while unsightly and vulnerable, greatly lowers the cost of deployment. This is particularly the case since the fiber cable medium is physically so much smaller and lightweight than other new services, such as coaxial cable, that might also compete to add to existing aerial communication and power facilities.

For aerial installation of FTTH facilities [Chomycz], one may either use **self-supporting** fiber cabling that actually contains its own strength member for

FIGURE 4.3 Street scene in a Tokyo residential area, illustrating the potential improvement to the visual environment promised by use of fiber's bandwidth for carrying multiple services.

FIGURE 4.4 Typical installation of lashed aerial fiber cable.

wind, icing, and other high-load conditions, or one might lash the fiber cable to a preexisting cable or a separate **messenger**, a preinstalled strength member. In the self-supporting case, no special equipment is needed for the span between poles, but when attaching to a messenger or preexisting cable, speedy installation is afforded by special **lashing** equipment, as shown in Figure 4.4. There, a truck with a cable reel feeds out the fiber cable to an overhead rope-towed or self-propelled lasher device containing a magazine of lashing wire that is spun out into a continuous helical wrap as the device moves along the supporting medium. As with underground installations, standards bodies have specified maximum pulling force allowed, 600 lb in the outdoor case and 300 lb in the indoor case [ANSI/TIA/EIA].

Special measures must always be taken to maintain or exceed the bend radius minima we shall discuss in Section 4.9, not only for the propagation reasons mentioned in Section 3.4 but also because excessive kinking, even momentarily, could sever one or more fibers in a bundle. Bend radius considerations also come into play when the fiber cable traverses a pole, since it is customary to leave slack at each such pole, either as a sag (catenary) or in the form of a simple "expansion loop," a coil of cable that reserves a meter or so of cable for future splice operations. In the first case, standard practice dictates that the length of the catenary should be at least twice the sag, which in turn should exceed the allowed bend radius.

While the most common means of overhead deployment involves either self-supported cabling or the use of lashing, hollow plastic ducts are occasionally used. These are more ubiquitous in underground installations, which we discuss next.

■ 4.4 UNDERGROUND DEPLOYMENT

With growing public reaction to the unsightliness of aerial utility lines, their physical vulnerability, and (for conductive such lines) their electrical vulnerability to lightning, underground deployment has become a growth industry. In many parts of the United States there are few new multiunit real estate developments that do not feature buried power and communications. And, here and there, communities are demanding that existing facilities that are aerial be moved underground.

Much ingenuity has gone into developing new methods of trenching and boring for placement of fiber bundles underground, and we shall now go through the wide range of challenges faced, and the solutions available, for underground FTTH deployment. Which method is chosen depends heavily on several factors:

- Whether the area of the installation is rural or urban
- Whether the path traverses paved areas
- Whether the area is subject to heavy mechanical traffic loads
- Whether penetration by ground water is likely
- Whether rocks in the soil might create pressure-induced puncture or unacceptably tight bending conditions
- Whether the area is subject to heavy damage by rodents or vegetation

This last factor can be a surprisingly important concern. In the Australian Outback, there is a parasitic shrub (the Christmas bush), whose underground tentacles seek out host roots from which to suck nourishment, and upon finding one, attach to it and then gradually sever it. The telephone company's clever solution was to encase the small cable in such a large tube that the parasite considered it to be some sort of pipe, not a root, and therefore too big to attack. In parts of the American Southwest, rodents can destroy any unprotected cables that are less than 6 ft below the surface [Kaminow].

The simplest installation method is **direct burial**, in which a trench is dug, the cable or duct laid, and the trench backfilled. If ducts are not used, the cable may be in direct contact with any rocks, roots, or other underground hazards, which, if not severing the fiber can often produce unacceptably tight bend radii.

Direct burial has been highly automated in recent years by the use of sophisticated plows, such as the one shown in Figure 4.5a, which cuts a trench in the soil in the form of a groove as deep as 2.1 m, deploys the cable or duct from a reel, and even backfills and smooths over the trenched region, all in one pass. Railroad rights-of-way are a favorite place to deploy intercity and some suburban telecommunications media, and special cable-laying railcars have been developed that hold reels of cable or duct, which are then deployed by plow mechanisms projecting sideways and downward from the vehicle, so that the cable is plowed in some several meters out from the roadbed.

If soil conditions require digging a narrow trench rather than plowing the cable into the ground, one can dig a wide trench by hand or by conventional excavation equipment, or use a dedicated **trencher**. The one shown in Figure 4.5b can dig a trench 9 to 24 in. wide, and as deep as 2.1 m, and then place the fiber within it.

Microtrenching is the name sometimes given to the technique, often used in built-up urban areas, of sawing a narrow slot across a paved surfaces using a

(a)

(b)

(c)

FIGURE 4.5 (*a*) Cable plow, (*b*) trencher, and (*c*) saw. All of these are interchangeable attachments for the same cable-laying machine [DitchWitch].

rotary saw, placing a modest cable bundle in the slot, and then backfilling with some mastic material. A typical saw for such a process is shown in Figure 4.5*c*.

One of the most remarkable recent developments has been **horizontal directional drilling (HDD)**, also known as **trenchless deployment**, which allows ducts or bundles of ducts to be placed at a fixed distance below the surface in a controlled azimuthal direction. Horizontal directional drilling originated in 1971 with river crossings for pipelines. The drill operator noticed that some of the drill heads he had earlier discarded because they did not drill a straight hole could be deliberately launched at a certain tilt and could be counted on to emerge near a predicted point on the other side of the river after having made a vertically curved path that stayed below the river bottom. Since then, monitoring of the drill's trajectory at the surface, and indeed a certain amount vertical and especially horizontal directional control have become routine.

A machine for this purpose is shown in Figure 4.6. An inner pipe carries a remotely powering downhole electric motor that drives a rotating bit that bores through dirt, a mixture of soil and stone, and even any hard and substantial rock encountered. A standard drilling fluid called bentonite is injected into the hole through this inner pipe and then carries out the debris through the outer pipe, which, as shown in Figure 4.6, comes in interconnecting sections stored on a rack atop the machine. The outer pipe thrusts the entire assembly forward and is capable of small changes in azimuthal direction of the hole, as commanded by the machine operator. Also at the end of the drill is a low-frequency radio transmitter (*beacon*), whose signals are monitored by the operator's monitor unit at the surface, in order to deduce the necessary depth and steering information. Depth is usually measured by knowing in advance what the receive beacon signal strength should be as a function of depth, while horizontal position is determined by measuring the receive signal difference between two directive receptors whose directivity maxima are offset by a slight angle. The beacon also transmits a data stream that includes information on roll, pitch, transmitter temperature, and status of its battery (if any). Small HDD rigs for FTTH service, such as the one shown in Figure 4.6, are capable of spanning 150 ft, a distance long enough for most driveways and small parking lots, whereas large machines can cover up to 1000 ft. Some years ago an HDD rig drilled a hole for telecomm facilities from the then World Trade Center in Manhattan under the Hudson River to a location in Jersey City.

FIGURE 4.6 Directional drilling machine [DitchWitch].

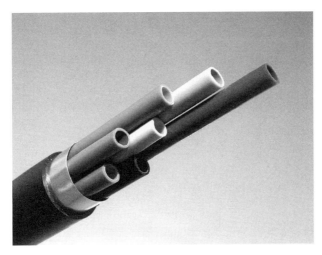

FIGURE 4.7 Plastic duct and subducts to contain optical fiber cabling.

Progress in cost-reducing fiber-laying technologies has progressed to the point that the actual cost of laying an underground FTTH network today ranges from $2 to $5 per foot using the combination of open trenching, plowing, and directional drilling that typical circumstances require. This is to be compared to $1.50 to $2.00 per foot for aerial deployment [Render].

Microtrenching, ductless direct burial, and directional boring do not by themselves allow later modifications to be made to the number and nature of the cabling that they carry. For decades, tunnels or sizable metallic ducts within buildings and under the streets have been the standard way for telephone, cable and power companies to deal with this flexibility problem. Now a miniaturized version of such solutions is available in the form of the plastic ducts and subducts, an example of which is shown in Figure 4.7. Subducts are typically available in sizes down to 5 to 8 mm, and ducts (bundles of subducts) are available in a wide variety. In the case of directional drilling, these ducts are pulled into the hole after the drilling is completed. They are often installed with the initial complement of fiber already in place inside. Or, analogously to **dark fiber** (installed, but as yet unused runs of fiber), it is now economical and practical to install **dark ducts**, empty, but usable in the future. It is nice to be able to have such a flexible "pay as you grow" facility, but if such fiber pathways are subject to change, how does one remove any existing fibers and insert new ones?

Removal is the simpler of the two problems. By having the duct or subduct interior surface suitably lubricated at manufacture, removing fiber bundles by pulling, while maintaining tension within the 600-lb standardized limit, is a soluble problem. As for installing the fiber, new cable is usually pulled down the duct behind a pilot length of strong pulling cable inserted previously. Today, the problem of future-proofing the installation, that is, facilitating insertion of new or replacement cabling, has been effectively dealt with by the invention in 1982

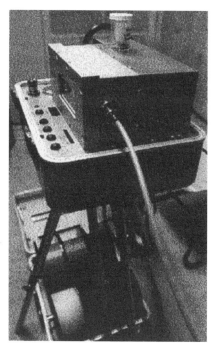

FIGURE 4.8 Typical fiber blowing system.

of **blown fiber** techniques in which the cable is pushed, not pulled, using airflow. A typical such mechanism is shown in Figure 4.8.

In earlier installations, the compressed air propelled a bobbin down the duct, which in turn pulled the fiber attached to it. It was then realized that the concentrated strain on the leading portion of the fiber could be reduced considerably by manufacturing the fiber jacket with dimples in it to transfer the airflow forces into a completely distributed pulling force on the fiber. Such dimpled fiber cable is now widely available. For five 5- to 8-mm inner-diameter subducts, up to 12 SMF or MMF fibers can be blown simultaneously down each one. Blowing distances between manholes or other points of access of up to 1000 m are now routinely available, and fiber installation can take as little as 90 min. Fiber can also be blown vertically, as is necessary in tall office or apartment buildings. Vertical blowing distances up to 300 m have been reported [Blolite]. One interesting wrinkle is that ducts and subducts can be pressurized so as to provide security alerts in case tampering or tapping is attempted.

■ 4.5 REUSE OF UNDERGROUND FACILITIES

In the early days of fiber optics, one of the largest startup carriers, Williams Communications, was built on the novel idea of buying up unused intercity oil pipelines, pumping out all the oil, and running optical fiber in the robust mechanical duct formed by the metal pipe. Today there is a local access echo of that early long-haul history in the increasing use of sewer and storm drain facilities in dense metropolitan areas. This is an environment where pathways for new communication facilities are extremely scarce—few or overcrowded telco or power company ducts, no overhead pole lines permitted, everything covered with concrete or occupied buildings, complex legal entanglements, and so forth.

In such desperate circumstances, recourse has sometimes been had to the reuse of water, gas, and especially sewer and rain runoff piping that has existed for some time, often for centuries. Special camera-bearing robots have been developed over the years to survey such facilities when they are too small for human inspection and even to make mechanical repairs. And then, with the growth of communication bandwidth, methods have recently been developed to extend this robot technology by causing it to place expandable **clip rings** at intervals inside the pipe, and then route a number of bundles of fiber along a series of hooks that protrude inward from each ring. The toxic and chemically destructive nature of sewage, and the absolute necessity of not allowing the fiber installation to impede fluid flow, either of sewage, drinking water, or storm water runoff, dictate that the fibers be suspended from the roof of the pipe.

■ 4.6 CABINETS, PEDESTALS, CLOSURES, AND VAULTS

As was shown in Figure 4.1, there are many points throughout the system where fiber cables must be terminated, spliced, or split, and these points may be placed within closures that are hung on the messenger cables or from self-supporting fiber transmission cables on poles, may be mounted within weatherproof small **pedestals** or larger cabinets long grade, or inside underground **vaults** below grade.

There is a very wide variety of shapes and sizes of suspended splice **closures**, several types being visible in the Tokyo street scene of Figure 4.3. In the United States, splice closures are standardized in Telcordia's GR-771 document [GR-771]. Most closures are of cylindrical shape, made of plastic, and can be split out lengthwise into two halves for installation, repairs, and alterations, as illustrated in Figure 4.9, which shows the housing opened out and the technician populating a **splice tray** with many mechanical splices. Moisture seal glands are provided at either end where the cables enter and leave. Capacities for housing splices or connector pairs typically range from 2 to 144 fibers, and these numbers match the commonly available numbers of fibers per cable. The availability of small, narrow planar waveguide splitters has encouraged deployers in Japan to frequently do the splitting up in the elevated closure, rather than in the ground-level cabinet or pedestal.

Fiber distribution hub cabinets too are available in a wide variety of sizes and capacities and are usually mounted on concrete pads at ground level, on utility poles, or in vaults. Figure 4.10 shows a typical unit. Because much of the system testing at installation time and for later troubleshooting is done at the ground level local convergence points or NAPs, connector pairs are often used in preference to splices, particularly downstream from the splitter.

FIGURE 4.9 Craftsman populating an aerial closure with mechanical splices. (Fiber cleaver at right, bottle of index-matching gel at left.) [3M]

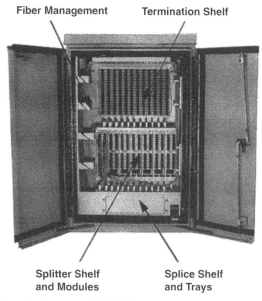

FIGURE 4.10 Fiber distribution hub [FONS].

The placement of fiber splicing and coupling facilities in underground vaults accessed by manholes poses special problems for the installer, for example, cramped working space, possible flooding, an occasional varmint, and the necessity of special ventilation during working periods. Nevertheless, the relative physical and electrical invulnerability and a negligible environmental impact often make this option a preferred one, especially in urban areas.

Multifiber cable with from 96 to 432 fibers commonly comes with every bundle of up to 32 fibers either as a ribbon or contained within a common plastic loose tube. In deploying a PON, it is common to route the entire cable to one ground-level fiber distribution point containing the splitters, and to terminate one bundle there. The unterminated bundles do not get cut but continue onward down the right-of-way, with the option of spatially reusing any of the fibers in the bundle that had already been terminated.

For the FTTH systems that are being deployed in the United States, the tedious splicing in of splitters and splicing out of drop cables are the last steps in deploying the ODN and are done as sets of customers are added. The single upstream fiber of a splitter is commonly fusion spliced, while the numerous downstream fibers from the same splitter may be connectorized, commonly using APC (angled) connectors, and without index-matching gel, for environmental reasons.

■ 4.7 SUBSCRIBER PREMISES OPTICAL NETWORK UNIT

The OLT and ONU are the demarcation points between fiber and copper at the head end and subscriber premises, respectively. An example of an ONU is shown in Figure 4.11*a*. Figure 4.11*b* shows a view of the various electrical and optical connections. A typical ONU might have four RJ-11 connections to four different 24-gauge twisted-pair phone connections (different phone numbers), RJ-9 connections for analog TV (54 to 890 MHz for up to 83 standard NTSC channels in the United States), and one RJ-45 connection for a CAT-5 line carrying high-speed Ethernet data (and voice if VoIP is used) to a small consumer-level Ethernet router elsewhere in the home.

Optical network units such as the one shown in Figure 4.11 are usually made with a double door. The inside one is available only to craft personnel, whereas the outer one may be opened by the subscriber for access to the premises connections only. With residences in the United States, the ONU is almost always placed outside so as not to impede access by craft personnel (either for tests and repair or to terminate service for nonpayment of arrears by the subscriber). In Japan, the ONU is usually indoors, which saves considerably in cost.

As mentioned earlier, it is important to minimize the per-subscriber technology cost—central office per-line costs are less important, especially with a PON, where the per-port cost is shared, typically 32 ways. The ONU embodies at the photonics level a low-cost Fabry-Perot laser diode for 1310 nm upstream, and a standard PINFET photodetector for 1490 nm downstream, a 1550-nm photodetector (if

(a)   (b)

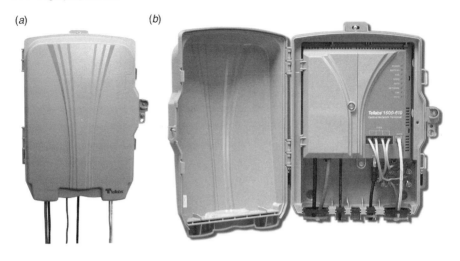

FIGURE 4.11 Optical network unit (ONU): (*a*) external view and (*b*) internal view of connections. (Copyright © 2005 Tellabs. All rights reserved.)

needed for analog TV), plus suitable optics for separately routing light of the three wavelengths.

For maximum economy of cost and space, it is common to integrate several of these functions in a single **triplexer** component on the circuit board inside the ONU. The triplexor shown in Figure 4.12*a* contains the 1310-nm laser diode and separate PINFETs at 1490 and 1550 nm. The latter must be highly linear as required for multichannel analog TV service. The coarse WDM function can be provided by several means, a typical one being as shown at Figure 4.12*b*. It uses a series of interference filters, thin-film devices, some placed at an angle, each of which passes certain wavelengths and rejects others. A similar set of optics can, in principle, be used at the transmitting end OLT, but with each light path reversed. When only the two wavelengths, 1490 and 1310 are used, as in Japan, the solution is a **diplexer**, simply two-thirds of a triplexer.

■ 4.8 HEAD-END OPTICAL LINE TERMINAL

Figure 4.13 shows the OLT unit deployed at the hub (head end or central office) of a typical PON. A fully populated such unit can support up to 22 PONs of up to 32 subscribers each, a total of 704. Back haul is handled by a DS3, OC3 or Gigabit Ethernet card.

Such units not only embody line cards to execute the BPON, GPON, or EPON protocols, but also support management and supervisory functions as well, usually

(a)

(b)

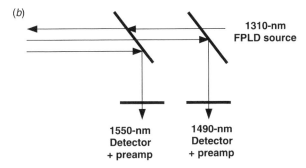

**1310-nm
FPLD source**

**1550-nm
Detector
+ preamp**

**1490-nm
Detector
+ preamp**

FIGURE 4.12 Typical 1550/1490/1310 nm triplexer [Infineon]: (*a*) overall device and (*b*) one way of configuring the optical paths.

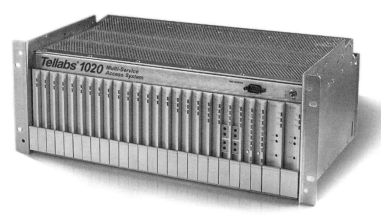

FIGURE 4.13 Head-end OLT unit supporting up to 704 subscribers. (Copyright 2005 ©
Tellabs. All rights reserved)

based on point-and-click **graphical user interfaces (GUIs)**. These include the usual diagnostics, as specified in the standards documents that were discussed in Chapter 2.

When 1550 nm is used for outbound multichannel analog TV, two unusual requirements must be met that are not involved in POTS, namely digital data or digital video service. As we saw in Chapter 2, higher power must be delivered from the OLT at this wavelength through the use of EDFAs. Moreover, extreme linearity is required in the modulation of the analog mix of many TV channels onto the laser's output, as we saw in Chapter 3. Both of these requirements add considerably to the cost of OLTs when analog video must be carried, and they also add somewhat to ONU costs.

■ 4.9   SLACK MANAGEMENT

Inside buildings or in the outside plant, in all pedestals, vaults, closures, cabinets, and terminal equipment of any kind, single-mode fiber's susceptibility to excessive losses due to bending must be taken into account. The loss occurs when the HE_{11} mode of Figure 3.3 is slightly distorted and begins to leak some energy into the cladding. In contrast to electrical wires and cable, where everything can be dressed down to exact lengths between connections with no slack remaining, every one of these pieces of single-mode fiber hardware, when properly designed, will be seen to need **slack management** racks or reels—some sort of fixtures that allow coiling of many centimeters of extra fiber. This not only allows the minimum bend radii to be maintained or exceeded, but it also provides extra lengths of fiber in case splicing or connector or splice replacement is required.

Gradual curvature of optical waveguides or fibers is a fact of life that electrical connections do not have to live with. (One very special exception is mentioned in the next section.) The LSI revolution in electronics finds but a pale reflection in most of the fiber-optics world for this reason. Certainly, splitters, certain photonic switches, and a few other optical components can be made smaller than they were several years ago, but the tyranny of the bend radius requirement will always mean that **photonic integrated circuits (PICs)**, such as the planar waveguide splitter of Figure 3.11, must be of fairly low density compared to the astonishing circuit density being achieved with electronic integrated circuits.

The permissible lower limits on radius of curvature of single-fiber and multifiber cable have been standardized [ANSI/TIA/EIA]. For single-fiber cable, the minimum is 2 in. when not under strain and 1 in. when being pulled. For multifiber cable, the minima are 10 times the outer diameter and 15 times, respectively.

■ 4.10   IN-BUILDING INSTALLATION

Figure 4.14 shows one commonly used nomenclature for the various portions of a premises fiber installation. There are horizontal runs in metallic or plastic ducts (or without ducts), vertical runs (**risers**), closets, and per-room outlets for cabling in a

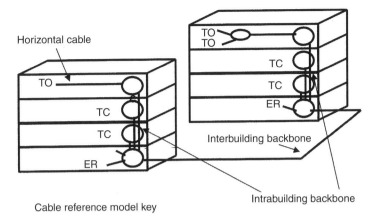

Cable reference model key

ER = Equipment room
TC = Telecommunication closet
TO = Telecommunication outlet

FIGURE 4.14 Nomenclature of in-building fiber [TIA/EIA-568A]: ER = equipment room, TC = telecomm closet, and TO = telecomm outlet.

building or group of buildings. The applicable standard for in-building installation and testing is that of the National Electrical Contractors Association and the Fiber Optic Association [NEC/FOA-301].

Thanks to the inventiveness of purveyors of fusion and mechanical splicers, connectors, and associated diagnostic equipment, in-house fiber is beginning to approach the low costs of in-house copper, but with the same low attenuation, low maintenance cost, and futureproofness that has become so attractive in the outside plant. One would think that as fiber costs drop and per-user bit rates grow, fiber might replace even such wideband copper media as Category 5 cabling and coax within the premises, even if the premises means not just a large apartment dwelling or office building but possibly an individual home.

With large MDUs and office buildings it already has. In such cases, the glass-copper boundary of the ONU can be placed nearer the user. Japanese experience has been that the last 100 m dominate the cost of the optical distribution network. NTT is therefore doing extensive research and development aimed at *do it yourself deployment.* For example, experimental "fiber to the desktop" systems are using special hole-assisted fiber for the last few meters from a connector in the wall to one in the set-top box or PC. A typical cross section of such fiber was shown in Figure 3.2c. The attenuation is insensitive to kinking and small bend radii, permitting bends with radii as small as 5 mm, as contrasted with the 2-in. lower limit for conventional SMF mentioned in the preceding section [Shinohara]. The disadvantage is that the attenuation is higher than that of conventional single-mode fiber, but since the runs are short, this makes little practical difference. One may expect much wider use of hole-assisted fiber in FTTH applications in the future, particularly on users' premises.

Multimode premises fiber is in wide use for local area networks (often to the desk), and it is ironic that such fiber, which is cheap and convenient to connectorize and splice, is not suitable for continuing an all-optical path from ONU to terminal equipment in an FTTH application—such fiber must be single mode. The reason is **modal noise**, which occurs when only one or a few permissible propagation modes of a multimode fiber are excited, and the delivered energy jumps around among these few modes, which are insufficiently numerous to give a smoothed average output power. At a point where single-mode fiber delivers a signal into a multimode fiber (e.g., by some kind of lensing system), essentially all the incoming energy will propagate at some mode or another in the multimode fiber. However, the converse is not true. In the reverse direction, the amount of HE_{11} energy that happens to arrive at the junction to the single-mode fiber from the multimode fiber, varies randomly with time.

Recall that modal noise is easily confused with mode partition noise, an effect observed in Fabry-Perot (F-P) lasers when the power output jumps around among the available F-P modes of the laser cavity.

■ 4.11 SAFETY CONSIDERATIONS

There are two potential safety problems that arise with FTTH systems that are unique to optical transmission system technology: damage to the eye's retina and ingestion of glass fragments. The nonlinear effects that can occur at high powers are not the only phenomena that limit the maximum optical power that should be carried on a fiber. There is also an upper limit on power above which it is possible to direct harmful levels of exposure into the eyes of humans who are using and maintaining the system. Fortunately, the longer wavelengths of FTTH systems, 1.3 to 1.55 μm, are much more highly attenuated in the outer layers of the eye than are any in the 450- to 700-nm band of visible wavelengths, whose attenuation in the cornea, lens, and eye interior are very low. One of the disadvantages of short-distance use of multimode fiber, driven by LEDs or **VCSEL (vertical cavity surface-emitting lasers)** is that the 850-nm wavelength at which the VCSELs are most practical is not that far below the long-wavelength cutoff of transmission into the retina, and therefore the eye safety limits for such multimode systems are much more aggressive.

Numerous standards bodies have set limits for the long wavelengths of PONs, for example, the American National Standards Institute (ANSI). ANSI's Z.136.2 document, as well as document 825-1 and -2 of the International Electrotechnical Commission (IEC) give the maximum power level numbers shown in Figure 4.15 [Bellsouth, Lucent] for classes 1 and 3. (There are also classes 2 and 4, but neither is relevant to FTTP since the former deals only with visible wavelengths and the latter with such things as laser welders, but not telecommunications.)

Upon comparing the numbers given in Chapters 2 and 3 for laser and EDFA power output at OLT or ONU, it is seen that, except for the EDFA power amplifiers used at 1550 for outbound analog video, most emitters fall within class 1 where the

Laser Class	Maximum Power Levels 1310 nm. 1550nm.		Safety Requirements
Class 1	8.8 mW	10 mW	Inherently Safe • Protective housing to prevent higher than classified emission • Safety interlock in housing to prevent access to nonclassified emission levels • Classification labels on the product • Caution lables on service panels, whether interlocked or not • User safety information in manuals
Class 3a	25 mW	50 mW	Safe Unless Viewing Aids Are Used Requirements in addition to above: • Key control • Beam stop to automatically disable the laser if access is required • Audible or visible "laser-on" warning
Class 3b	500 mW	500 mW	Additional Requirements • Remote control switch to allow disabling laser by door circuit • Aperture lable to indicate location of radiation output

FIGURE 4.15 IEC 825-1 and 825-2 classes of lasers, power limits and safety requirements [Bellsouth].

only safety precautions necessary are safety labels and downward-pointing protective housing for the laser or LED. Only those EDFAs of output power exceeding 10 mW will require class 3a or 3b special treatment, such as automatic shutoff in case of fiber break (where it is unknown who might try to look at the broken fiber end), visible laser-on warnings and physical access controlled behind lock and key.

By now, most technicians know that looking into the end of the operating fiber ("with one's remaining eye") is dangerous, and that it is fortunately no longer necessary to examine the fiber under a microscope because portable microscope/ video display units are available. An example is shown in Figure 4.16.

As for the problem of fiber fragments, they are essentially glass needles, which can be quite painful if stuck in one's skin and even life-threatening if ingested. The only formal standard [NECA/FOA-301] that addresses this problem simply sets several commonsense rules, including maintaining cleanliness of the workplace and also keeping a few patches of masking tape handy with which to make sure that all fiber cuttings are disposed of, including patting down clothing with such adhesive material upon completion of the job.

■ 4.12 POWERING

As we have seen, although some fiber installations have metallic strength members to ease the mechanical loading of the fiber and its various claddings and

FIGURE 4.16 Hand-held video microscope for viewing facet of a fiber, for example, one already deployed in a connector [Cabletesting].

jackets, basically an FTTH system is not one in which there is a natural way to power the ONU from the head end. By contrast, the important power supply capability of legacy copper-based systems, whether telco twisted-pair or cable provider coax, has played a significant role in various arguments in the public utilities commissions and with the public as to whether FTTH providers have an obligation to continue this service. In most parts of the United States, when there is a natural calamity such as a hurricane or earthquake, the POTS resources will be the last to lose power, and people have gotten used to this reassuring contact with the rest of the world when disaster hits. The installation, care, and feeding of a large central office 48-V battery banks, backed up by motor generator equipment that can keep power on the system as long as fuel is available, has constituted a major investment overhead for telephone companies worldwide.

The wave of common carrier deregulation and increased competitiveness that began in the 1980s has weakened this commitment to centralized power by telcos and their cable competitors and has caused both industries to be increasingly satisfied with localized rather than centralized subscriber battery backup. For telco subscribers, power unreliability has been exacerbated by the increased use of fiber-fed **subscriber loop carrier (SLC)** units, handling entire neighborhoods, with each SLC supplied with only a few hours of its own battery backup. The decreasing emphasis on telco responsibility to provide centralized power is also driven by the proliferation of cell phones, but their base stations often have only battery backup, and less frequently have automatically actuated motor–generator sets. One way or another, the public seems either to be unaware of the problem or is getting used to the idea that wired and cellular phone power may fail.

In this environment, FTTH providers clearly feel that they have a responsibility to provide an **uninterruptible power supply (UPS)** but not a responsibility to provide duration of continued service of more than a few hours, typically 6 to 8.

The battery UPS at an individual residence can either be inside or outside an exterior wall, just as the ONU itself can be at either place. While the inside installation will be cheaper, this advantage can be offset by the increased difficulty and cost of service calls such as battery replacement, which require the involvement of the occupant. For such reasons, outside installation seemed to be the preferred solution today, at least in the United States. In Japan, there are experiments with giving the subscriber the responsibility to replace ONU power, which can take the form of ordinary flashlight batteries.

One UPS solution is shown in Figure 4.17. The ONU is powered by the UPS battery, which is floated across a small DC supply driven by 110/220-V power from the adjacent electric meter. The higher cost of such an outdoor installation, compared to an equivalent indoor version, is partly due to the increased environmental protection required, in particular an electric heater for maintaining battery viability in cold weather.

At the head end one may need either 120- or 220-V wall plug AC or 48- or 60- or 90-V DC power, or both AC and DC. Either way, all the DC systems in the center must be supplied by a battery UPS, and the AC systems with an additional battery-driven DC-to-AC motor generator. Usually such battery facilities are in turn backed

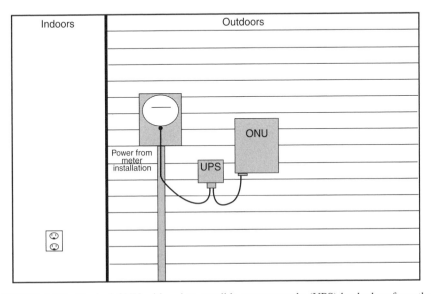

FIGURE 4.17 Typical ONU with uninterruptible power supply (UPS) backed up from the electrical power meter.

up by a diesel motor-generator set. The batteries may be required to carry the load for only a few minutes, time enough for the motor-generator set to start up.

The evolution of FTTH powering has been complicated by the fact that wall plug power is 110 or 220, most premises ONUs use 12 V DC, cable companies distribute power at 60 to 90 V DC, and the standard POTS supply voltage has always been 48 V DC. This is somewhat reminiscent of the proliferation of many other kinds of telecommunications standards.

■ 4.13 TESTING AND MAINTENANCE

By their very passive nature and topological simplicity, the P2P and P2MP optical access systems have lower costs of network installation, management, and maintenance than do the older copper-based networks. However, there are a number of special optical signal quality tests that must be made during the installation, when services are activated, and in troubleshooting an already activated system.

The test equipment for this work typically includes:

- An **optical loss test set (OLTS)**, embodying a source of light and an optical power level meter
- An **optical time-domain reflectometer (OTDR)**, a sort of optical fiber radar set that emits a short pulse into the fiber and displays the strength of the returned echo
- A special microscope–video display unit (Fig. 4.16) for inspecting fiber ends and connector facets

The first two instruments must be capable of testing the system at all the wavelengths in use, 1310, 1490, and also 1550 nm if analog TV is involved. In fact, testing at this wavelength is particularly crucial because of the sensitivity of analog TV to out-of-spec signal levels, reflections, and other impairments.

During construction and installation, the system is routinely tested for attenuation by comparing a pair of power level readings using two OLTSs, one at the beginning and the other at the end of the fiber segment as the network is being built, or spanning much or all of the PON tree as it is being completed. This ensures that fiber kinking, bad connectors or splices, or a fiber break will not cause the overall loss budget to lie outside the range stipulated in the standard (Chapter 2). This measurement is made in both directions and is typically done by two OLTS units that have been calibrated against each other. The next step is to measure the **optical return loss (ORL)**, the ratio of incident power to reflected power at an input, expressed in decibels. If the ORL is too low, reflections causing this condition can produce fluctuations in laser-transmitted power (increased RIN), and double reflections can cause the receiver to experience undesired interference components that elevate the bit error rate in digital receivers and cause even more harm with analog receivers.

The ORL measurements must be made in both directions using either an OLTS or OTDR at one end and then the other. When undesired echoes are detected beyond one of the splitters, in order to identify the guilty party, loss measurements out to each subscriber are needed. One clever scheme for accomplishing this from the OLT proposes to install a reflective fiber Bragg grating (FBG) (Section 3.2.1) at each ONU, each such FBG having been tuned to a different signature wavelength. Tuning the OTDR to each subscriber's FBG wavelength then diagnoses attenuation to that subscriber [Merel].

The OTDR is a particularly effective instrument for uncovering pathologies of PON systems. It is capable of seeing any kind of distributed or discrete points of loss or reflection and spotting its location. Figure 4.18*b* shows the kind of trace that can be seen on such an instrument for the topology of Figure 4.18*a*. After each transmitted pulse (typically a few microseconds long), there is received a slowly declining ramp in the reflected signal due to the backward-traveling Rayleigh scattering. Any significant spatially compact loss, such as that from a connector pair, splice, or

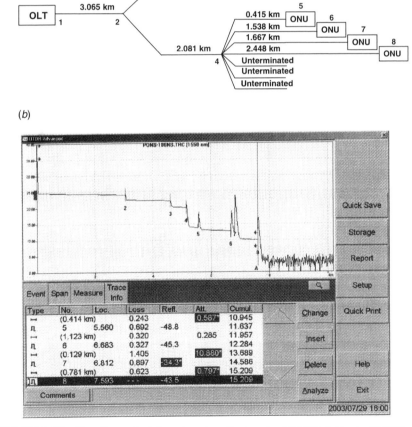

FIGURE 4.18 Use of an optical time-domain reflectometer (OTDR): (*a*) typical PON layout and (*b*) resulting OTDR trace [Chabot].

splitter, appears as a drop in the returned Rayleigh scatter, and conversely, a reflection appears as a positive excursion.

Far and away the commonest pathology in single-mode fiber systems concerns connectors, particularly the possible presence of small submicroscopic particles on a facet or a possible scratch or dent. As we saw earlier, even a few tenths of a decibel of connector loss can lead to a conflagration where high-power EDFAs are used, and therefore for some distance downstream in a PON delivering 1550-nm video only fusion splices are permitted. The ability to inspect for small particles with video microscopes has now progressed out of the laboratory and into the field, as we saw with Figure 4.16.

■ 4.14 COSTS

All of the various components of an FTTH installation that we have discussed in this chapter have been subject to cost improvements as well as economies of scale, as the FTTH solution to the broadband access bottleneck has become more widely adopted. For example, in the United States, per-home FTTH costs have dropped in 12 years by a factor of almost 5, as shown in Figure 4.19.

This is for homes actually connected and receiving service, not just homes passed. The $1650 per home for the year 2004 breaks down as follows: $250 for the per-subscriber fraction of the OLT at the head end, $400 for each ONU, $375 for the prorated cost of the ODN, the glass network between head end and subscriber, and $625 for engineering and constructing the ODN. The $250 for head-end cost does not include any analog video equipment.

As for the individual optoelectronic components, Ryan, Hanken, and Kent give the following for 2005 costs in quantity: an OLT transceiver costs $160, triplexers for the ONU cost $75 to $100 in the United States, and the diplexors (1490 and 1310 nm) used in Japan cost $60. Splitters are down to $13 per port. For 1310-nm

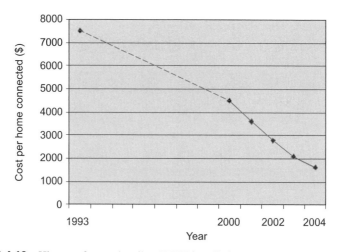

FIGURE 4.19 History of per-subscriber FTTH installation costs [Render].

gigabit Ethernet (or EPON upstream) service, directly modulated lasers cost $22, and APDs $100 [RHK].

As one might expect, the largest cost component lies in the construction process. Aerial construction can be as low as $1.50 to $2.00 per foot, while underground costs per foot range from $2.00 for plowing to $12 for directional drilling. When streets must be dug up, as is sometimes the case in high-density areas, costs may range from $15 to $30 per foot [Render].

We see from Figure 4.19 that, on the average, FTTH cost per home has been dropping by a factor of roughly 0.75 annually. The Moore's law curve of the cost of semiconductor-based electronics as a function of time for the same performance has been dropping by about the same factor [Desurvire]. The ability of FTTH cost reduction to come anywhere close to the celebrated Moore's law curve is remarkable, considering that only about one-third of the $1650 per home cited above is electronics. This speaks well of the innovative capability of the civil engineering profession when confronted with the new challenges of deploying optical fiber.

REFERENCES

[3M] www.3m.com/intl/kr/network/pdf/2178_LS_series.pdf.

[ANSI/TIA/EIA] American National Standards Institute, Telecomm. Industry Assoc., and Electrical Industries Assoc., Standard 568B.3.

[ANSI] American National Standards Institute and National Electric Contractors Association, Standard NEIS-301, 1998.

[Barany] D. Barany, B. Murphy, and R. Draper, Various Powering Methods for FTTX Architectures, undated.

[Blolite] www.blolite.net.

[Bellsouth] Technical Reference 73603, Unbundled Dark Fiber (UDF) Technical Specifications, February, 1999.

[Cabletesting] www.Cabletesting.com.

[Chabot] S. Chabot and M. LeBlanc, Fiber-Optic Testing Challenges in Point-to-Multipoint PON Testing, 2003 Fiber to the Home Conference, New Orleans, Paper TO3.

[Chomycz] B. Chomycz, *Fiber Optic Installer's Field Manual*, McGraw-Hill, New York, 2000.

[Corning] www.corning.com/deployment\ftthome.html.

[DitchWitch] Model RT 185/HT185 attachments. www.ditchwitch.com. See also Product Literature CD, The Charles Machine Works. Perry, OK.

[Desurvire] E. Desurvire, *Global Telecommunications—Broadband Access, Optical Components and Networks, and Cryptography*, Wiley Interscience, Hoboken, NJ, 2004, p. 301.

[FONS] www.fons.com/documents/FDH_Data_Sheet.pdf.

[Gariepy] D. Gariepy, M. Leblanc, B. Masson, and B. Ruchet, FTTP and PONs—Where Does Testing Come in? 2004 Fiber to the Home Conference, New Orleans, Paper TO4.

[George] J. George, Designing Passive Optical Networks for Cost Effective Triple Play Support, 2004 Fiber to the Home Conference, New Orleans, Paper TO5.

[Girard] A. Girard, Passive Optical Networks—Critical Issues Related to Technology and Testing, 2004 Fiber to the Home Conference, New Orleans, Paper F12.

[GR-771] Generic Requirements for Fiber Optic Splice Enclosures, Teleordia Technologies, August, 1994.

[IEEE] Institute of Electrical and Electronics Engineers, Standard 802.3ah, 2004.

[Infineon] Tri Port-BIDI®, www.infineon.com.

[Jefferies] FTTx and Current Trends in Outside Plant Infrastructure Millenium Skyline Project Report, New York, Aug. 11, 2004.

[Jensen] R. Jensen, Why Fiber Links Fail, Cabletesting.com, 2002.

[Kaminow] S. E. Miller and I. P. Kaminow, *Optical Fiber Telecommunications, Vol. II*, Academic Press, 1988, Chapter 5.

[Lucent] Technical Note, Laser Safety and Optical-Fiber Communication Systems, Microelectronics Group, Lucent Technologies, March, 1999.

[Merel] R. Merel, private communication, February, 2005.

[NECA/FOA-301] Installing and Testing Fiber Optic Cables, National Electrical Contractors Assoc. and Fiber Optic Assoc., 1998.

[Poynton] C. Poynton, *Digital Video and HDTV*, Morgan Kaufman, San Francisco, 2003.

[Render] Render, Vanderslice and Associates, *Fiber to the Home—The Third Network*, Tulsa, December, 2004.

[RHK] 2005 cost data from K. Liu, Ryan, Hanken and Kent, March 2005.

[Shinohara] H. Shinohara, Broadband Expansion in Japan, Plenary presentation, OFC/NFOEC Conference, Anaheim, March 8, 2005.

[TIA/EIA 568.A] Commercial Building Wiring Standard, Telecomm Industry Assoc. and Electrical Industries Assoc., Jan. 1995.

[TIA/EIA 568.B.3] Optical Fiber Cabling Standard, Telecomm. Industry Assoc. and Electrical Industries Assoc., April, 2000.

a VOCABULARY QUIZ

Discuss not only what these terms abbreviate but also what they mean.

Aerial deployment	Distribution cable	Messenger	SLC
ANSI	Duct	Microtrenching	Splice tray
ASE	Extinction ratio	Modal Noise	Splitting loss
Beacon	FDH	MTU	Terminals
Bit error rate	Feeder	NAP	Thermal noise
Bitrate	GUI	OLTS	Trencher
Blade	HDD	ORL	Trenchless
Blown fiber	IEC	OTDR	deployment
Cabinet	Index matching gel	Patch panel	Triplexer
Clip rings	Lashing	Pedestal	UPS
Closure	LCP	PIC	Vault
Dark ducts	Line card	Risers	VCSEL
Dark fiber	Link budget	Self-supporting	
Diplexer	Loss budget	cable	
Direct burial	MDU	Slack management	

CHAPTER 5

Current Deployments

■ 5.1 INTRODUCTION

Different countries are responding differently to the imperatives we sketched out earlier and the availability of highly evolved FTTH technical solutions. In some countries geography dictates a need for solutions providing long reach as much as for high bit rates—much more than can be provided by DSL (Figs. 1.7 and 1.8), while in others the distances are much smaller. In some countries, HDTV looms as a major emerging requirement, while in others it is not legal to bundle broadcast television service into a mix that also includes POTS or data. Some countries have a tradition of government funding of new technological directions of promise to the society, while other governments find such top-down stimulus to be anathema, either because of risk aversion or due to the imperatives of a favored ideology. In some countries the driver is cable competition; in others competition from DSL, and in still others international competition.

These differences between national cultures lead to a striking disparity in the degree to which developed countries have embraced broadband in general and FTTH in particular. By broadband we mean cable modems, DSL, or FTTx. Nation-by-nation differences are evident from Figure 5.1, which shows the per-capita availability of broadband of all forms. It is seen that penetration is strongest in Asia, with Europe in second place and North America not even on the charts. This relative ranking is also shown in Figure 5.2, which gives FTTH penetration in terms of numbers of lines rather than the per-capita uptake of Figure 5.1.

We now shall discuss in more detail the state of FTTH deployment in different parts of the globe.

■ 5.2 UNITED STATES

There are 176 million telco voice-grade access lines in the United States today, of which 75% run all the way from the CO and 25% from some form of digital loop

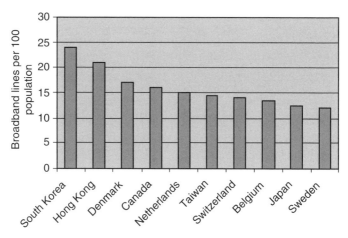

FIGURE 5.1 Per-capita broadband availability in various countries as of second quarter of 2004 [Point-topic].

carrier node somewhere between the CO and the subscriber. One would think that the prevalence of longer access distances (Figure 1.2) in North America would cause FTTH deployment to be more advanced there than elsewhere in the world. One would also have expected FTTH deployment not just by small companies in small communities, as has been happening for some years, but also as a major effort by the major players, the incumbent local exchange carrier (ILEC). The ILECs are Verizon, Bell South, SBC, and Qwest, often referred to as regional Bell operating companies (RBOCs).

To take another look at some matters that were discussed in Chapter 1, there were several reasons that deployment by the North American ILECs began only very recently:

- Excessive state and federal regulation, a heritage from times past when land-line communication was a monopoly. This has been an inheritance worldwide

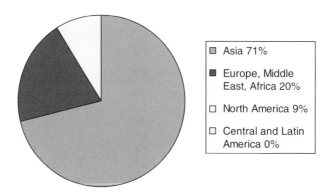

FIGURE 5.2 FTTH lines by region [Whitman].

but somehow seems to have become more of a deterrent in the United States than elsewhere.

- As part of the root cause of this regulation, a strong tendency on the part of the North American ILECs to strive for competitive advantage not by technical or market innovation but by case-by-case legal action and by lobbying efforts for favorable state and federal legislation and court rulings.

- The fact that, until very recently, while overlapping geographical coverage existed for the long-haul carriers (AT&T, Sprint, MCI), it did not for the local exchange carriers. This led to a modus vivendi of de facto noncompetition among the ILECs.

- A technical preoccupation with continued incremental improvement of a technology (DSL) that exploits what is probably the ILECs' strongest asset—the installed base of millions of miles of twisted-pair copper—but is one of the weakest answers to future broadband requirements.

- The lack, until recently, of any serious competition (other than wireless) for high-capacity broadband last-mile service.

- The lack, until recently, of the bit rate usage drivers mentioned in Chapter 1, especially high-performance laptop and desktop computers and also HDTV.

As Figure 5.3 shows, the FTTH that has been deployed has come in the form of a first wave of small fragmented implementations, with a larger second wave of major ILEC deployment. The second wave is just beginning, due so far solely to the recent initiatives of the largest ILEC, Verizon.

By the spring of 2005, the first-wave companies had passed some 200,000 homes, MDUs, and small businesses in 398 communities and actually connected over 80,000 of them. In a major turn-around, Verizon, starting at the beginning of 2004, passed 1.6 million premises by spring 2005 and expects to pass 3 million

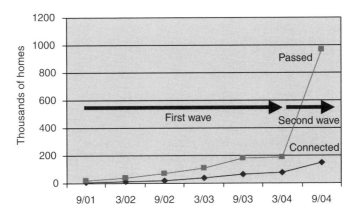

FIGURE 5.3 Results of a U.S. "census" on percentage installed FTTH homes passed by provider segment. The first wave is due to small companies, and the second wave mostly due to one ILEC, Verizon [Render].

by year end 2005. By April, 2005, the number of homes that Verizon had actually connected and were producing revenue was over 200,000 [Render]. It should be noted that Verizon still aggressively markets new DSL service, even in its own FTTH service areas.

Most first-wave installations were carried out by startups with less than 150 employees each and are taking place in small, scattered places that are not always considered priority markets for either the large telcos or large cable companies. Interestingly, this pattern in which a new technology is first introduced in small communities is just the way the United States cable television industry got started, and for many of the same reasons. The first-wave FTTH deployments in the United States are approximately evenly split between APON and EPON architectures.

The first-wave deployments in the United States have been done typically by:

- *New greenfield housing developers* There are 1 to 2 million new housing starts per year in the United States, and new homes add 1 to 2% to the existing population of access telecommunication facilities per year. This has been the largest immediate market sector for the first-wave FTTH system providers. The initial cost of laying fiber is essentially the same as for coax. When it costs the installer only 1000 to 2000 extra dollars in FTTH equipment costs to help sell a house costing up to hundreds of thousands of dollars, developers use FTTH as a significant purchase incentive for the new owner.

- *Municipalities* Over 1800 small communities own their own electric utilities and therefore have the required rights-of-way and the utility poles already at hand. Such communities are helped economically by the availability of subsidies from the Universal Service Fund (USF), the pool of federal money created decades ago to which incumbent carriers must contribute in order to ensure that rural and small town citizens not be disadvantaged.

- *Publicly owned electric utilities* There are at least 70 sizable public utility districts nationwide, serving a total of more than 1.5 million customers. Also, into this category one should probably lump the many rural coops that pool resources to buy and resell power or communication services. USF subsidies are available for them as well. As with municipality owned power facilities, fiber resources are sometimes already in place to monitor the power distribution systems, sometimes all the way to the premises electric meter.

- *Competitive local exchange carriers (CLECs) and overbuilders* Often, small CLECs will "overbuild" into the territory of neighboring large ILECs and take business away from them on the basis of the claim that they can provide better or cheaper service than the resources on top of which they overbuild.

- *Small ILECs* Most of the 1300 or so small independent telcos are heavily committed to DSL, at least for the moment.

Until 2003, ILEC deployments had been limited to temporary and inconclusive, single-community demonstrations, in a pattern that has become very familiar to telco watchers over the years. However, in June, 2003, three of the four

ILECs got together to issue more or less identical **requests for price quotation (RPQs)** based on a common BPON standard. This focused attention on FTTH as a significant business opportunity for many responding small companies, large companies, and consortia. As a consequence great attention was given by the economically suffering telecom components industry to questions of costs, standards, and optimization of the technologies for FTTH, rather than for metro or long haul service.

Thus, unlike many other such ILEC initiatives through the years, the 2003 RPQ led not simply to a series of tentative field trials but to the serious entry of Verizon, and thereby the arrival of what we have called here the second wave. As of mid-2005, they were building FTTH facilities in 12 states. As the statistics of Figure 5.3 shows, in 1 year the number of homes passed in the nine initial Verizon suburban service areas has far exceeded the cumulative total passed by all of the small first-wave companies in their 398 communities.

Verizon and the other ILECs chose the BPON form of ATM PON, with stated intention to go to GPON within several years. This preference is based partly on the telcos' traditional concentration on ATM rather than Ethernet, on inherited use of circuits rather than packets, and can also be partly blamed on the lateness of publication of the formal EPON standard. Now that publication has occurred, EPONs are being seriously considered by Verizon and the other ILECs.

One should note that, so far, all the numbers shown in Figure 5.3 are miniscule compared to the 2004 figures of 7.1 million DSL and 14.8 million cable modem U.S. broadband customers. However, the significance of the FTTH numbers lies in the rate of growth, the fact that 4 years ago these numbers were all close to zero, and the fact that FTTH is now expected by many to expand rapidly, now that at least one ILEC seems to have made a serious commitment.

The Verizon service offering, called **FiOS (fiber-optic service)**, typically offers to residences guaranteed bit rates per user of 5 Mb/s downstream and 2 Mb/s upstream for $40 per month or 15/2 for $45 or 30/5 for $200. There is also a 5-Mb/s symmetrical service for businesses. These services are currently based on that BPON version (Chapter 2) that specifies 622 Mb/s downstream and 155 upstream. Video, when made available is handled by RF carrier at 1550 nm. There is a stated intention to go to 100 shared Mb/s in the future.

Verizon spokesmen quote several reasons that they expect their business to be much more efficiently managed when the current transformations built around FTTH are completed [Lacouture, O'Byrne]. These include a number of reasons having nothing to do with simply capacity improvements:

- Manual order taking is replaced by Web-based ordering.
- Allocation of copper pairs is replaced by allocation of pieces of bandwidth.
- Service activation is done by software rather than by a truck roll.
- Today's limited fault isolation capability is upgraded with centralized online performance monitoring.

Demographically, this second wave of FTTH penetration by Verizon focuses on relatively prosperous suburban areas of large cities, as contrasted with the first wave, which tended to involve smaller communities that are less a part of an extended metropolitan area.

In the emerging competition for triple-play services that the telcos face from the cable companies, there is some consolation from the fact that so far the cable companies have very little penetration of businesses relative to their penetration of the much larger residence market. The decades-long relationships that the ILECs have built up with businesses, large and small, will likely prove to be a key advantage.

The other U.S. ILECs have basically adopted different strategies. SBC has limited its FTTH installations to greenfield cases (new builds), relying on ADSL2+ copper tributaries for most of its broadband customers. With its **Lightspeed** service, it intends, by 2008, to overbuild with this FTTC service to 19 million homes (roughly half their total), and to provide FTTH service over PONs to 1 million new homes. SBC is rapidly moving to an all-IP backbone using 10-Gb/s Ethernet and IP routers [Wallace]. It is working with Microsoft toward making IPTV the workhorse TV delivery method.

Bell South appears committed to fiber to the curb, possibly because so many of its access lines are buried, whereas Verizon has the advantage of having the largest percentage of aerial access lines of any of the ILECs. It was Bell South who successfully challenged in court the FCC's insistence on all-fiber access as the price of regulatory forbearance. The resultant ruling says that, as long as the DSL copper in an FTTC installation extends no farther than 500 yards, the same relief from the **unbundling** requirement that had already been afforded full FTTH is available. This was one of the deregulatory factors that assisted the move to FTTx in the United States. Bell South's plan is to start with ADSL, then go to ADSL2+, followed by VDSL as the bit rate requirements evolve, particularly those from HDTV [Fahmy].

■ 5.3 JAPAN

Japan leads the world in number of subscribers served by FTTH, even though its populated areas are among the densest, allowing the competing ADSL form of broadband to achieve very high penetration and aggressive bit rates. Whereas in the United States competition from cable is the principal driver, in Japan competition from DSL acts as a similar stimulus.

The Japanese success with FTTH lies especially rooted in traditional Japanese ways of launching new industries based on new technologies. A top-down national consensus seems more easily formed there than in places like the United States. There is often widespread coordination between corporations in one narrow new business areas in the same industry, while aggressively competing in others. Some new industrial initiatives succeed and some do not. Few remember the unsuccessful "third-generation computer" initiative of the 1970s, which was intended to put Japan at the top of the large general-purpose and scientific computer industries, with a heavy dose of artificial intelligence. On the other hand, it is equally easy to forget that the term "VLSI" first gained prominence in the Japanese national initiative of the same name, which really did succeed and propelled Japan into the leadership position, particularly with memory chips.

Today, partly thanks to another such top-down initiative, known as **e-Japan**, the country is enjoying a leadership position in FTTH, with, for example, significant demand for Ethernet-based PONs at gigabit rates. e-Japan was initiated in 2001 with the goal of making Japan the most advanced nation in this part of information technology by 2006. Proposed targets included using fiber to interconnect 30 million subscribers at 10 Mb/s rates and 10 million subscribers at 100 Mb/s rates, all of this by year 2010 [Shinohara, Hanatani].

By August, 2004, there were 1.6 million FTTH subscribers, and new ones were being added at a rate of 100,000 per month [Sato]. This is to be compared to 12 million DSL users and 2.7 million cable modem users. Most of the FTTH systems have been BPONs. For example, the dominant communications carrier, **Nippon Telephone and Telegraph (NTT)**, has widely deployed two-wavelength BPONs that use 1550 μm downstream rather than the 1490 μm stipulated in the standard that was described in Chapter 2. This is because NTT, being legally prohibited from delivering broadcast TV, has no way of using 1550 for analog video, with its attendant difficulties with EDFAs and the nonlinear effects that ensue. Small satellite dishes seem to be the preferred way of receiving broadcast TV in Japan. Of course,1550 is the most desirable wavelength band from the attenuation standpoint.

NTT's FTTH service, called B-FLETS (B for broadband), already provides residents with 622 shared Mb/s downstream and 155 upstream and provides businesses 100 Mb/s in both directions. The B-FLETS service is moving off of BPONs and onto EPONs that have a shared bit rate of 1 Gb/s symmetrical (see Chapter 2). Such symmetric Ethernet-based PONs at gigabit rates are called GEPONs in Japan, to emphasize the gigabit rate, but inviting confusion with ITU's ATM-based GPON. The per-subscriber bit rate of GEPON service is 100 Mb/s on a best-effort basis. As of mid-2005 GEPON deployment had already begun in the NTT-East service area. NTT is technically a holding company that controls NTT-East, NTT-West, NTT-Comm (mobile data services), NTT-Data (data services to businesses), and NTT-DoCoMo (mobile radio).

It is unclear whether the GEPON is part of an aggressive national program to get rid of the 125-μs framing of all the legacy telco architectures and to move onto an Ethernet base for everything, including long haul. If this does take place, it may happen slowly, since NTT aggressively markets not only long-line Ethernet but such other services as primary rate ISDN, frame relay, and ATM [Shinohara].

In addition to NTT, a number of other competing providers are active in Japan [Whitman, Converge]. Tepco offers 1 Gb/s service to multidwelling units. KDDI (Kokusai Denshin Denwa International), formerly the undersea cable company, is also active with FTTH, as is the cable TV company J-Comm. SoftbankBB has started offering GEPON service in Tokyo. Even the U.S.-derived Yahoo!BB, a large Japanese DSL provider, is entering the competition for FTTH services by leasing NTT lines. USEN specializes in serving condominiums in 11 Japanese cities [Americasnetwork]. In addition to these services, a consortium involving Poweredcom, Toshiba, and Tokyo Power Company has recently announced that it will start using the FTTH facilities of others to offer DVD downloads with a form of copy protection. These are to be available on either a sales or a rental basis. Remote file backup has also been spoken of as a possible service offering [Render].

■ 5.4 KOREA

Korea can boast the largest per-capita penetration of broadband anywhere, 11.6 million subscribers, 79% of the households, plus 24,000 Internet cafes. There is also a significant cable modem presence. The **VISION-2006** residential broadband initiative of the Ministry of Information and Communication has a target of 100 Mb/s to 5 million subscribers by 2007 and 10 million by 2010. They are noncommital as to technology choice, but these aggressive bit rates are likely to dictate some sort of PON.

On the other hand, the Electronics and Telecommunications Research Institute has had since 1982 a series of pilot studies and field tests of FTTC and FTTH, either passive P2P or P2MP or active (with electronics between CO and subscriber) [Song]. They appear to be fully prepared for large-scale deployment when the time is right. In addition to these architectures, the Institute has been experimenting with WDM PONs using various methods of overcoming the cost barrier of selecting and stocking laser diodes for many specific wavelengths. They have tried slicing the spectrum of a high-power LED by passing it through a channel-specific narrow-band filter and have also experimented with injection locking of Fabry-Perot laser diodes [Lee].

■ 5.5 CHINA

The People's Republic of China had over 24 million broadband subscribers by mid-2005, the largest number in the world. All of this is DSL today, but there are field trials and a vague plan for eventual migration to FTTH. Fifteen million new DSL lines are being added per year. The largest carrier is China Telecom, with China Netcom (CNC) in second place. Both have declared the intention to offer FTTH at some time in the future [Whitman].

■ 5.6 AUSTRALIA

There is a scattering of greenfield FTTH deployments in new Australian residential developments. In the city of Ballarat, FTTH deployment began in mid-2004, offering IPTV, VoIP telephony, as well as data. In Queensland, the largest national carrier, Telstra, is conducting FTTH PON field trials [Whitman].

■ 5.7 EUROPE

By June, 2004, there were roughly 547,900 FTTH subscribers in Europe and 1.96 million homes passed [Whitman]. (As throughout this book, we have been including MDUs and small businesses in our definition of FTTH). Most of these systems, as elsewhere in the world are supplied with rings upstream from the OLT. The leaders are Sweden, Italy, Denmark, and Netherlands.

By mid-2005, Sweden had 200,000 subscribers in citywide FTTH systems in 290 municipalities. Bredband offers 10 Mb/s bidirectional service to "fiber sockets" in apartments and is said to have 150,000 paying subscribers. The take rate is very high, 39%—that is, 39% of the homes passed were connected.

In Italy, with 190,000 FTTH subscribers and a take rate of 15%, one provider, Fastweb, has the FTTH game almost entirely to itself. Its FTTH offering features 10 Mb/s symmetric data plus POTS, video, and video on demand. Its subscribers [Whitman] are concentrated in 13 cities, where it also supplies some DSL.

Denmark has the unbelievable take rate of 76%. Netherlands, although one of the most compact countries in Europe, has a take rate of 67%.

Most of the FTTH installations in Europe are undertakings of public institutions, typically municipalities interested in direct business development advantages or the advantages of providing superior living situations. The lack of PTT involvement is reflected in the almost complete lack of BPON/GPON penetration relative to EPON.

One can see differences in the reported country-by-country rankings on FTTH penetration, for example, in comparing the Q2 2004 data of Figure 5.1 with the European numbers just cited. However, the data all show that FTTH is advancing rapidly in Europe.

REFERENCES

[Americasnetwork] www.americasnetwork.com/americasnetwork/article/articleDetail.jsp? id=93689.

[Converge] www.convergedigest.com/DWDM/DWDMarticle.asp?ID=12512.

[Fahmy] H. Fahmy (Bell South), presentation at OFC/NFOEC Conference, Session NTu1, Anaheim, March 8, 2005.

[Hanatani] S. Hanatani and K. Nishide, Fiber to the Home in Japan and Comparison with the U.S., Presented at organizing meeting of Fiber to the Home Council-Asia, Tokyo, Oct. 19, 2004.

[Johnson] T. Johnson, Point-Topic, Ltd., private communication, November 2005.

[Lacouture] P. A. Lacouture, Keynote talk at 2004 Fiber to the Home Conference, New Orleans, October 6, 2004.

[Lee] S.-M. Lee, et al., Dense WDM-PON Based on Wavelength-Locked Fabry-Perot Lasers, Poster paper JWA55, OFC/NFOEC Conference, Anaheim, March 9, 2005.

[O'Byrne] V. O'Byrne (Verizon), presentation at OFC/NFOEC conference, Anaheim, Session NTu1, March 8, 2005.

[Point-topic] www.point-topic.com.

[Render] Render Vanderslice and Associates, *Fiber to the Home—The Third Network*, Tulsa, December 2004.

[Sato] K. Sato, Prospects and Challenges of Photonic IP Networks, presented at Asia-Pacific Optical Communications Conference (APOC), Beijing, November 2004.

[Shinohara] H. Shinohara, Broadband Expansion in Japan, Keynote presentation at OFC/ NFOEC conference, Anaheim, Session NTu1, March 8, 2005.

[Song] H. Song, Broadband in Korea, October 19, 2004 presentation.

[Wallace] M. Wallace, Project Lightspeed, OFC/NFOEC conference, Session NTu1, Anaheim, March 8, 2005.

[Whitman] R. Whitman, International FTTH Deployments—Lessons Learned Around the Globe, FFFH Council Annual Meeting, New Orleans, October 6, 2004.

a VOCABULARY TEST

Discuss not only what these terms abbreviate but also what they mean.

e-Japan	Lightspeed	RBOCs	VISION-2006
FiOS	MITI	RPQ	
IPTV	NTT	Unbundling	

CHAPTER 6

The Future

In the first chapter we outlined some of the pressures from customers, from governments, and from competitors that are causing providers of last-mile solutions to move away from copper and into the potent and essentially future-proof newer technology of fiber. Chapter 2 told how the required architectures and protocols have reached maturity. In Chapter 3, the various component technologies involved were examined, and it was pointed out here and there that the special imperatives imposed by FTTH, particularly on cost reduction, have already borne fruit. Chapter 4 told a parallel story about a different technology sector, that of the deployment process. The general thrust of these overviews has been that FTTH is a very mature way of doing things, both technologically and economically. Certainly the overall cost improvement in both base technology and deployment technology has been significant and can be expected to continue.

Unfortunately, while FTTH is technically mature, it is not very mature yet as a part of twenty-first-century society. However, things are getting better. For example, if this story had been told as recently as 2 years ago, one would have had to say that in the United States the only parts of the culture that were taking FTTH seriously enough to invest heavily in it were small companies and small communities. This has now changed dramatically.

The subtitle of this book is "The New Empowerment," and the preceding chapters should have made it clear that proliferation of fiber to the home and business promises an enormous empowerment to all the elements of society, and not just the end users. New application directions will be ignited and will create new jobs in all businesses, especially within the telecomm and computer industries. FTTH also offers the providers of this new access technology further business and management freedoms from the tyrannies imposed upon them by their nineteenth-century copper and circuit-switched technology and its regulatory encrustations. Furthermore, the simplicity and flexibility of fiber-based systems, and above all their lower lifetime costs, should accelerate the penetration of broadband across to the less fortunate inhabitants of the bottom side of the digital divide.

By supporting new jobs and new businesses opportunities where remoteness is no impediment, FTTH can contribute greatly to the economic recovery of rural America.

The substitution of fiber for century-old copper with its inherited traditions, particularly its regulatory and competitive traditions, amounts to a clean slate on which new regulations, to the extent that many are required, can be applied in an up-to-date way. However, there is always somewhere a societal down side to any revolutionary technological advance, and FTTH is no exception. In particular, the regulators will have to understand and deal with the fact that such stupendous last-mile capacity in the hands of the first provider to deploy FTTH will give that provider such a head start that some new thinking about monopoly power in this setting will be absolutely required. After all, the up side of the Tokyo street scene of Figure 4.3 is that a number of telecom providers were making a living off each of the parallel and obviously duplicative facilities.

Not only will traffic growth in the long-haul and metro do a lot to revive the entire telecomm industry, but this will once again lead to bandwidth starvation in those sectors. Dense WDM (and some content of photonic switching) will rise from the grave. For years, many of us were predicting "all-optical" networks— those in which there are no active components between end users. However, it seems we were coming at this problem from the wrong direction. Instead of starting where the traffic intensity is the greatest, we should have started where the bottleneck is the tightest. This is neither in the metro nor the long-haul, but in the access—the last mile.

Now, as the preceding pages have indicated, we have come full circle from the predictions of a decade ago—we have an increasingly deployed all-optical network, but guess what? Instead of this all-optical network being the predicted continental and intercontinental complex of dense WDM nodes with fully photonic switching, we have a simple, passive, and localized but extremely effective solution, the P2MP PON, a truly primitive all-optical network.

It is not at all clear today whether, as time passes and the insatiable human greed for more and cheaper bandwidth persists, the entire path between subscribers will become all-optical. Certainly massive photonic switching and massive WDM can handle the traffic, but the required header recognition and processing when the signal is a stream of photons and they are all IP, not circuit switching, is today an unsolved problem.

Index

Fiber to the Home: The New Empowerment, by Paul E. Green, Jr.
Copyright © 2006 John Wiley & Sons, Inc.